Technology in the Country House

Technology in the
Country House

Marilyn Palmer and Ian West

 Historic England

 National
Trust

Published by Historic England, The Engine House, Fire Fly Avenue, Swindon SN2 2EH
www.HistoricEngland.org.uk

Historic England is a Government service championing England's heritage and giving expert, constructive advice; the English Heritage Trust is a charity caring for the National Heritage Collection of more than 400 historic properties and their collections; the National Trust is a charity that looks after places of historic interest or natural beauty permanently for the benefit of the nation across England, Wales and Northern Ireland.

The views expressed in this book are those of the authors and not necessarily those of Historic England.

First published 2016

ISBN 978-1-84802-280-5

British Library Cataloguing in Publication data
A CIP catalogue record for this book is available from the British Library.

The right of Marilyn Palmer and Ian West to be identified as authors of this work has been asserted by them in accordance with the Copyright, Designs and Patents Act 1988.

Historic England holds an unparalleled archive of 12 million photographs, drawings, reports and publications on England's places. It is one of the largest archives in the UK, the biggest dedicated to the historic environment, and a priceless resource for anyone interested in England's buildings, archaeology, landscape and social history. Viewed collectively, its photographic collections document the changing face of England from the 1850s to the present day. It is a treasure trove that helps us understand and interpret the past, informs the present and assists with future management and appreciation of the historic environment.

For more information about images from the Archive, contact Archives Services Team, Historic England, The Engine House, Fire Fly Avenue, Swindon SN2 2EH; telephone (01793) 414600.

Brought to publication by Victoria Trainor, Publishing, Historic England.
Typeset in Georgia Pro Light 9.5pt
Edited by Kathryn Glendenning
Indexed by Caroline Jones
Page layout by Pauline Hull
Printed in the UK by Bell & Bain Ltd

Front cover
A Buzaglo stove, originally installed in the Great Hall at Knole but now in the Orangery.

Frontispiece
The electrical switch room at Castle Drogo.

Back cover
The gas 'Sun Light' in the Yellow Drawing Room, Wimpole Hall.

CONTENTS

FOREWORD

Britain's country houses and their estates are a potent legacy of a time when the country's rural landscape, and the lives of almost everyone in it, was dominated by wealthy private landowners. We are fortunate that so many of these properties not only survive but can be visited by the general public, affording an insight into this lost world. However, whereas previous generations of visitors generally went to marvel at the taste and refinement of their owners, and to enjoy the works of art and fine furniture which they collected, today's visitors are often equally interested in how these curious institutions operated, and in gaining an understanding of the parallel lives of the owning families and the vast armies of staff who served them.

Technology played an increasingly important role in enabling owners of country houses to achieve their ambition of a comfortable house which functioned efficiently and, as far as the owners and their guests were concerned, largely invisibly. This book shows, for example, how mechanical and then electrical bells were used by the household to summon servants from their distant quarters, whilst lifts and sometimes even railways conveyed fuel, food or luggage unseen to where they were needed. Perhaps even more significantly, this book explores the role that many country house owners played in the improvement and acceptance of many of the home comforts which we, in the developed world at least, take for granted, such as central heating, sanitation, running water and electric lighting. Other landowners, however, remained attached to the traditional ways of running their households and the book explores the question of how far the availability of servants before the First World War cushioned them against the need for change. The dwindling financial resources which many country estates experienced from the late 19th century has meant that the physical evidence of these earliest examples of domestic innovations has often survived rather than being swept away by later modernisation, as was often the case in town houses and middle class dwellings.

Archaeologists Marilyn Palmer and Ian West have spent several years exploring the attics, cellars and other hidden corners of almost a hundred country houses and their immediate surroundings all over Britain, not just those run by the National Trust and English Heritage but also some which remain in private hands. This has enabled the authors to compile this unique account of how, particularly after 1750, new technology shaped the development of country houses and their estates, and affected the lives of the people who lived and worked in them. They have benefitted greatly from the cooperation of those responsible for the curation and interpretation of the houses and have, in return, shared their findings with them. More and more houses are opening up new areas to the public – basements, cellars, bathrooms, service corridors and tunnels, estate yards and stables – with the result that visitors are now able to enjoy much more extensive visits to country houses and their estates than in the past. The detailed research which underpins this book is proving hugely valuable in raising our understanding of how country houses worked, and it is hoped that it will encourage even more people to explore these hitherto hidden facets of country house life.

It is very appropriate that this book should be published in collaboration between Historic England and the National Trust. As two organisations with a common interest in helping people understand and appreciate the historic environment, it is sensible to combine our educational and publishing expertise in this way, and we plan to co-operate on similar ventures in future.

Dame Helen Ghosh
Director-General
National Trust

Duncan Wilson
Chief Executive
Historic England

PREFACE

This book is dedicated to the memory of the late Dr Nigel Seeley, formerly the National Trust's Surveyor of Conservation. Recognising both a growing interest among visitors in the functioning of country houses 'below stairs', and a lack of understanding among property staff of the ways in which domestic services contributed to this, in 1996 Nigel instigated a rapid survey of all National Trust properties, advised by the National Trust's Industrial Archaeology Panel. The aim of the survey was 'to identify and record all early mechanical, electrical, gas and water systems in historic houses and their immediate curtilage, whether or not they are currently in use, and any other items of technological interest'.[1] Much of the work was conducted by Paul Thomas, a postgraduate student at University College London, where in 2002 Nigel took up a Visiting Professorship. Also working on the survey were Maureen Dillon, studying the artefacts of historic lighting for a doctorate, and Frank Ferris, a member of the Heritage Group of the Chartered Institution of Building Services Engineers (CIBSE), who visited many properties to survey heating equipment, publishing details on the CIBSE website.[2]

Nigel died in June 2004 and a symposium on country house technology was held in his memory at the Royal Institution in London in May the following year. Both this and the continuation of Nigel's work in this area were strongly supported by Sarah Staniforth, then the National Trust's Historic Properties Director, and managed by Linda Bullock and Helen Lloyd, the Trust's Preventive Conservation Advisers. Subsequently, Marilyn Palmer, who had been a long-standing member of the Trust's Industrial Archaeology Advisory Group and then a member of its Archaeology Panel, established the Country House Technology Project in 2008, based at the University of Leicester's School of Archaeology and Ancient History. She secured some funding from the Leverhulme Trust and recruited Ian West, one of her former PhD students, to assist her. This new project, although making use of the National Trust's Country House Technology database as well as reports and articles written by Paul Thomas, Maureen Dillon and Frank Ferris, was not limited to the Trust's houses but included those owned by English Heritage as well as private houses, some not normally open to the public. The project team has also undertaken more detailed analysis of the technology at several English Heritage properties, at the request of Andrew Hann, Properties Historians' Team Leader.

Some aspects of historic domestic technology have been studied by previous researchers and writers. The book that stimulated Marilyn Palmer's initial interest in country house technology was Mark Girouard's *Life in the English Country House: A Social and Architectural History*, which was one of the first books on country houses to recognise the importance of understanding how families and their employees and dependents used the houses that they built, and included a section on early country house technology.[3] Clive Aslet, too, in his book *The Last Country Houses* and his updated edition of this, *The Edwardian Country House: A Social and Architectural History*, recognised the growing importance of technological innovation in improving domestic comfort in the early 20th century.[4] His book included the first illustration the authors had seen of a basement vacuum cleaner. In the field of lighting, Leeds City Art Galleries published in 1992 a technological overview and catalogue of their exhibition, Country House Lighting, as part of the *Temple Newsam Country House Studies* series.[5] This was followed a decade later by Maureen Dillon's book *Artificial Sunshine*, based on her doctoral study of country house lighting.[6] Pamela Sambrook has written a considerable number of books on the subject of servants and their workplaces, including *A Country House at Work*, which examines the service areas at Dunham Massey, Greater Manchester, in some detail, together with a definitive study on *Country House Brewing in England, 1500–1900*.[7] Jeremy Musson's

Up and Down Stairs: The History of the Country House Servant is a comprehensive study of the working lives of servants into the 20th century.[8] Among a number of works focusing on the development of cooking appliances is one written jointly by Pamela Sambrook and Peter Brears, *The Country House Kitchen 1650–1900,* and one edited by Ivan Day, *Over a Red-Hot Stove: Essays in Early Cooking Technology.*[9] Several books from Shire Publications cover aspects of domestic and estate technology, including David Eveleigh's *Firegrates and Kitchen Ranges* and his *Privies and Water Closets,* together with volumes on garden technology such as Fiona Grant's *Glasshouses* and Rosalind Hopwood's *Fountains and Water Features.*[10] These and other similar publications have deliberately concentrated on specific areas of work in the country house and its estate but a broader review of country house technology as a whole has been provided by Christina Hardyment's *From Mangle to Microwave and Home Comforts: A History of Domestic Arrangements.*[11]

The Country House Technology Project has sought to build on and update the work of these previous researchers, highlighting in particular the impact which these innovations had on the planning of the country house, the factors influencing country house owners in whether or not to adopt new technology, and the effect of this technology on the lives of the people living and working within the house and its estate. One of the first outcomes of this project was a weekend conference, held at Rewley House, Oxford, in 2010, with contributions from some of the writers mentioned above, as well as the authors of this work; the proceedings of this were published in 2012.[12]

The authors' approach to this subject is unashamedly archaeological, driven, in the first instance, by the surviving remains of different technologies in country houses and their parks, gardens and estates. In the course of this work, it has become clear that the most reliable guide to the adoption of particular innovations is the physical evidence: plans may have been prepared and estimates provided by contractors, but that does not mean that the work was executed as indicated. The 19th century in particular was a period of fervent invention, much of it devoted to the improvement of home comfort and convenience; the pages of magazines such as *The Engineer* and *Scientific American* were full of ideas which never saw the light of day beyond the inventors' own homes.[13] Total reliance on printed sources such as journal articles and patents can result in a much distorted view of the history of technology. Consequently, the starting point for this study has always been the site visit to properties, almost a hundred of which have been made, covering all parts of the United Kingdom. Of course, some documentary research was undertaken prior to each visit to identify what to look for, but care has been taken to avoid visiting only houses known to have remarkable examples of domestic technology: much can be learned from those properties which failed to innovate, for whatever reason.

However, some caution has had to be exercised when examining the physical remains of technology in country houses, as discussed more fully in the conclusion to this book. Numerous examples were found of former gas lights converted to electricity (or even oil) in houses which never had a gas supply, the fittings having been introduced when electricity was first installed or, in some cases, as part of a later 'restoration'. Notable also was the property which had an electric bell indicator board in the corridor outside the butler's pantry but which had neither any bell pushes or bell wiring connected to it anywhere in the house – clearly the result of some well-intentioned attempt at authenticity after the house was taken over by the heritage body concerned. Surviving artefacts had to be considered within their wider context, both inside the house and across the estate, and documentary sources consulted to identify and date the appearance of features connected, for example, with water supply, energy production and sewage disposal, to support the interpretation of the physical remains.

The results of these site visits and follow-up research have been collated and analysed here in separate chapters for each broad area of technology, with the important interactions between different technologies highlighted. The authors make no apology for analysing some technologies in more detail than others: as indicated above, some aspects of domestic services, notably lighting and cooking, have received some attention in previous studies whereas very little has been published on other aspects of the country house, such as communications and heating. Some of the chapters also show how technology contributed to the self-sufficiency of country house estates, a theme which is further considered in the modern context in the conclusion to the book.

Marilyn Palmer and Ian West

ACKNOWLEDGEMENTS

During the course of the project which led to the writing of this book, we visited almost a hundred country houses around the United Kingdom. We received unstinting assistance from staff, volunteers and, on a few occasions, house owners, who took us to places within the houses and their estates which the public, and sometimes our guides, too, rarely see. It is impossible to list all of these people here but we thank them for their time and forbearance – every visit has added to our knowledge and understanding of the impact of technology on country houses in some way. We would, however, like to highlight and express our gratitude to a few such individuals, who have shown particular interest in our work and have provided continuing support and information as our research has proceeded: Robin Harcourt-Williams, former Archivist at Hatfield House; Christine Hiskey, Archivist at Holkham Hall; Paul Holden, House and Collections Manager at Lanhydrock; Chris Hunwick, Archivist at Alnwick Castle; Robin Wright, Engineering Curator at Cragside; David Good, owner of Myra Castle, and his son Michael Good; and Caroline Magnus, owner of Stokesay Court. Grateful thanks are also due to the countless staff at local records offices and specialist archives who have assisted us. The planning of our visits to National Trust properties was greatly aided by the surveys of domestic technology previously carried out by Paul Thomas and Maureen Dillon.

While the broad scope of our research generally deterred us from investigating the history of individual houses in minute detail, we greatly appreciated the opportunities provided by Andrew Hann, Properties Historians' Team Leader at English Heritage, which allowed us to conduct in-depth studies at Audley End and Eltham Palace. Both of these houses, in very different ways, were at the forefront of innovations in comfort and convenience. We are also very grateful to those experts whose knowledge of various aspects of domestic technology has been invaluable, including Frank Ferris and Paul Yunnie (Chartered Institution of Building Services Engineers, Heritage Group), Mick Kernan (Fire Service College), Professor Miles Lewis (Faculty of Architecture, Building and Planning, University of Melbourne), David Eveleigh (Director of Collections & Learning, Ironbridge Gorge Museum Trust) and Pete Smith (formerly of English Heritage). We would also like thank: the Leverhulme Foundation for providing some of the initial funding for the project; Paul Barnwell for reading some of the text; and Bill Barksfield, Managing Director of the tour company Heritage of Industry (www.heritageofindustry.co.uk), for permission to use a number of his photographs taken during specialist tours of country houses that we led.

Thanks are also due to those who saw the book through publication from start to finish including current and former National Trust personnel: David Adshead, Claire Forbes, Helen Lloyd, John Stachiewicz and Sarah Staniforth, and from Historic England publishing: John Hudson, Robin Taylor and Victoria Trainor, and also Kathryn Glendenning, for copyediting the text, Pauline Hull, for the design layout, and Caroline Jones for the index.

The National Trust gratefully acknowledges a generous bequest from the late Mr and Mrs Kenneth Levy that has supported the cost of preparing this book.

Introduction: the background to technological change in country houses

'Convenience, not symmetry, is now the universally acknowledged rule in house planning.'
J J Stevenson, 1880[1]

The United Kingdom's country houses comprise one of its greatest assets, not just as popular destinations for both domestic and overseas visitors but also as microcosms of rural life until at least the outbreak of the First World War. Although the composition of the landowning class changed dramatically in the course of the 19th century, the possession of a large house in the country remained the aspiration of old and new families alike. For many, the ownership of land itself was no longer of prime consideration, particularly after the fall of rents following the agricultural depression of the final decades of the 19th century. Many established landowners had long supplemented their agricultural incomes with, for example, profits from mining or from urban properties, while new entrants to the ranks of the landed gentry usually had incomes from trade, commerce or industry. What they wanted was a house in which to entertain their friends or, in some cases, their commercial clients, and most of them were prepared to spend money on making their houses as comfortable and welcoming as possible – as did Baron Ferdinand de Rothschild at Waddesdon Manor, Buckinghamshire (Fig 1.1).

In the course of the 18th and 19th centuries, various technological innovations became available to enhance the comfort and convenience of domestic life. Water supply had been improved in many country houses from at least the 18th century onwards, partly for the lakes and fountains popular in gardens of the period. If water was piped to the house, it was often only to the basement and ground floor, although there were some exceptions. Hot water supplies came much later, often only towards the end of the 19th century, and so fitted bathrooms with baths and washbasins were a comparatively late development. A good cold water supply was essential for improvements in sanitation, particularly water closets, early versions of which were installed in some houses from the 18th century onwards and went through many changes in design before the late 19th century. Country houses, usually being far from urban centres, had to deal with their own sewage, which was often discharged into rivers to start with, but more sophisticated treatment methods were developed by the 19th century.

Distance from towns also prevented country house owners from making use of urban gas supplies, which rapidly became available from the early 19th century, and they were forced to build their own gasworks and gasholders if they wanted to adopt this new energy source. Many eventually did, but quite slowly compared with their enthusiastic reception of electricity from the early 1880s, which again had mostly to be generated within the grounds of the country house; some country house owners were very much pioneers in this field.

Central heating systems became available from the early 19th century and were enthusiastically taken up in public institutions and industrial complexes, but much more slowly in country houses, where open fires reigned supreme. This was partly because it was felt that open fires were far better for ventilation and that breathing warm air was bad for health. Steam and hot water heating systems were, however, extensively used in greenhouses and other structures in kitchen gardens, often some time before they were used within the house. The increasing use of coal rather than wood for open fires as well as heating systems meant that country house owners had to think about ways of transporting coal to and even within their houses, and some made use of small railways for this purpose as well as, rather later, lifts driven first by hydraulic power and then by electricity. These lifts were also

Facing page
The Gold Medal Eagle range in the kitchen at Uppark, West Sussex. The range was installed c 1895.

used for the transport of food as well as people, although the rope-hauled dumb waiter was a more common method of moving food from basement kitchens to upstairs dining rooms.

As country houses grew in size, particularly with the construction of much larger service areas from the 18th century onwards, better methods of communication between the household and their servants had to be found. Sprung bells, worked by a complex system of wires, were introduced from the mid-18th century onwards and often remained in use until the 20th century. From the 1860s, electric bells began to replace or supplement these, driven by batteries as their installation often preceded that of electricity. Neither sprung nor electric bells, of course, enabled direct communication between the two groups, and speaking tubes were sometimes added to the communications system in the second half of the 19th century. These, however, had a limited range and were replaced by internal telephone systems from the 1880s onwards, although it was not until the 20th century that many houses were able to make use of external telephones, again because they were often too far from public telephone exchanges.

Fire was an ever present hazard in country houses and many owners provided themselves with the means of dealing with it, ranging from hand-operated fire pumps to sophisticated systems of fire hydrants, as well as making use of fireproof materials in any rebuilding that was done. Kitchens had often been placed away from the main house because of fire, but this had the serious disadvantage that food had to be taken quite a distance to dining rooms; means of reheating it in a nearby servery were introduced in the second half of the 19th century. Many new devices for cooking food also became available, notably better methods of turning spits for roasting in front of open ranges and the introduction of closed stoves which greatly increased the range of cooking methods available to the kitchen staff. Conversely, keeping food cool was revolutionised by the introduction of refrigeration from the end of the 19th century, supplementing the use of ice from the ice houses to be found on many country estates.

Other domestic activities benefited from technological innovation, notably the laundry, where heated drying cupboards and stoves for flat irons speeded up the process to some extent, although washing by hand continued well into the 20th century. Security was another concern, improved locks and safes being among the additional precautions which were introduced. Finally, by the

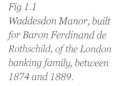

Fig 1.1
Waddesdon Manor, built for Baron Ferdinand de Rothschild, of the London banking family, between 1874 and 1889.

end of the 19th century, the vacuum cleaner was developed to alleviate the drudgery of the daily use of the dustpan and brush wielded by so many housemaids, particularly the powered basement vacuum cleaners which began to make their appearance in the first decade of the 20th century.

The pace of change was so great during the 20th century that the evidence for many of these early forms of domestic technology was swept away in most types of buildings. Since, however, country house owners had been obliged to expend considerable effort and expense in order to install these facilities in their rural retreats, modernisation was often much slower than in their town houses, particularly as many of them were suffering reduced incomes towards the end of the 19th century and could not afford any further changes. Consequently, the remains of the technologies which enabled country house owners to provide themselves with increased comfort and convenience – and which in some cases were pioneered by country house owners – are often still visible both in the houses themselves and on their estates.

Despite the loss of so many country houses in the last century, a sufficient number survive to demonstrate a past way of life. As Mark Girouard pointed out in 1978, 'Even when the customs have gone, the houses remain, enriched by the accumulated alterations, and often accumulated contents of several centuries. Abandoned lifestyles can be disinterred from them in much the same way as from the layers of an archaeological dig.'[2] The purpose of this book is to use exactly that approach to discover what evidence has survived of the impact of technological innovation on the building fabric, contents and landscape of country houses. Many books have been devoted to the life of those in domestic service in such houses. Rather than concentrate on the social records of their lives, however, this book aims to look at the archaeological and artefactual evidence for the increased comfort and convenience sought in country houses from the 18th to the early 20th centuries.[3]

Country house planning

In 1880, the architect J J Stevenson wrote in his book *House Architecture*, 'Convenience, not symmetry, is now the universally acknowledged rule in house planning.'[4] The terms 'comfort' and 'convenience' appear frequently in 19th-century books on country house planning. Robert Kerr, a Scottish architect working in London, produced one of the first of these books in 1864, *The Gentleman's House or, How to Plan English Residences from the Parsonage to the Palace*, which went through many editions. His definition of comfort was rather vague: 'Comfort includes the idea that every room in the house, according to its purpose, shall be for that purpose so contrived as to be free from awkwardness, inconvenience and inappropriateness.'[5] This is very similar to his definition of 'convenience': 'that characteristic which results from an arrangement of the various departments, and their various component parts, in such relation to each other as shall enable all the uses and purposes of the establishment to be carried on in perfect harmony – a place for everything and everything in its place'.[6] Comfort and convenience were to be achieved by careful planning and Kerr did not deal in any detail with technological innovations such as heating or sanitation. Stevenson paid more attention to the latter, but still argued that careful planning was the key, laying particular emphasis on what he described as 'multifariousness'. He stated, 'Keeping pace with our more complicated ways of living, we have not only increased the number of rooms, but assigned to each a special use.'[7] This can, of course, be seen from the plans of many country houses in the 19th century and will be considered more fully later in this chapter.

Gradually, though, the term 'convenience' began to take on more of its modern meaning. Lawrence Weaver, the Architectural Editor of *Country Life*, for example, referred to the telephone in 1911 as 'one of the most usual modern conveniences found in the country house'.[8] Conveniences of the kinds described in this book could create country houses that were more comfortable for daily living, but they were introduced within the context of the careful planning of rooms for specific purposes that dominated the thinking of country house architects in the second half of the 19th century.

However, the first half of the 18th century had witnessed the dominance of Palladianism, an architectural style deriving from classical prototypes, by means of which the aristocracy perhaps hoped to demonstrate the ancient virtue of the political system which they themselves controlled.[9] Such structures did not lend themselves easily to the placing of servants' quarters. In some houses, the service area was in the basement, with a spinal corridor connecting the various rooms and giving access to stairs at each end up to the main floors of the house.[10] The architects of some Palladian houses, however, such as Nostell Priory in Yorkshire, Kedleston Hall in Derbyshire and Holkham Hall in Norfolk, designed a main block and linked pavilions, with the service areas housed in one of the pavilions. This removed kitchen odours from the main house but had the disadvantage that food had to be carried long distances to the dining room. It did, however, mean that servants were no longer dug into a pit in the ground, as Girouard expressed it so succinctly.[11] By the end of the century, the idea of the *piano nobile* – that is, having the main family rooms on the first floor – had gone out of fashion as house owners wanted direct access through doors at ground level to their newly landscaped gardens and parks. Separate service wings at ground level were now possible as symmetry was no longer the principal design characteristic. Occasionally, existing structures could be adapted to serve as a service wing, as at Tredegar House, Newport, where one wing, probably part of the original medieval house, was converted into a massive servants' hall and other service rooms in the late 18th century (Fig 1.2). Even earlier is the enormous detached service wing at Petworth House, West Sussex, built in the 1750s after a fire in the original kitchen destroyed the main staircase

and persuaded the 2nd Earl of Egremont to banish the domestic offices from the house altogether. Similarly, James Wyatt added a service wing to Matthew Brettingham's Gunton Hall, the Norfolk home of Sir Harbord Harbord, in the 1770s, which was nearly as big as the main house and was described by contemporaries as the finest in the kingdom.[12] Such separate service wings were made possible by the increasingly widespread use of sprung bells, which enabled servants to be summoned from a distance now that they were no longer immediately visible to the household (*see* Chapter 7).

James Wyatt was very concerned with the layout of service areas; it has been suggested that 'well-laid-out sets of service areas with specific functions were a feature of Wyatt's house plans from early in the 1770s and 1780s, foreshadowing the complex service arrangements of 19th-century country houses'.[13] Kerr certainly stressed in 1864 that more attention had long been given to the layout of what he described as 'the Offices' than to the family's parts of the house.[14]

Country house plans of the early to mid-19th century demonstrate very clearly the principle of 'multifariousness' admired by both Kerr and Stevenson (Fig 1.3). This was intended to ensure that servants did not get in each other's way and that each had enough space to carry out their

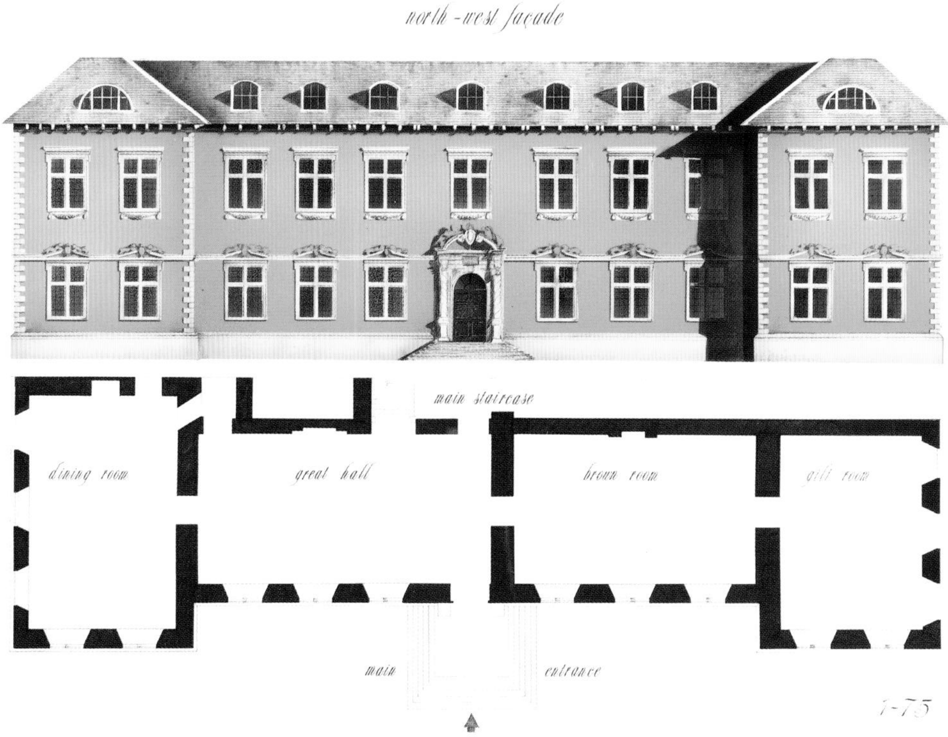

Fig 1.2
Elevation and plan of the servants' hall, Tredegar House, converted from an earlier medieval wing in the late 18th century.

Fig 1.3 (facing page)
Ground-floor plan of Buchanan House, Stirlingshire (William Burn, 1851–3). The service area, with its many separate rooms, is larger than the family area. From M Girouard, The Victorian Country House, *1976, Fig 2, p 32.*

north-west façade

main staircase

dining room   great hall   brown room   gilt room

main entrance

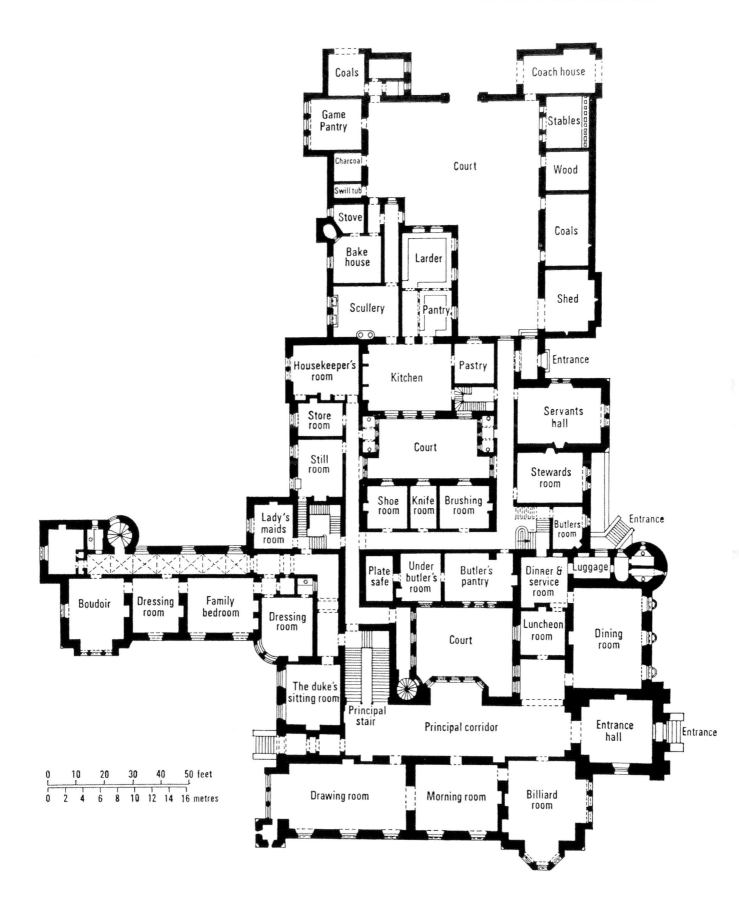

duties. Consequently, service areas were frequently a warren of corridors leading to small rooms, each of which had a particular purpose. The kitchen offices included: the main kitchen, with its cooking ranges; a scullery, not just for washing plates but also for the preparation of vegetables; a pastry kitchen away from the heat of the main kitchen; a still room for specialist preparation of jams, jellies and making tea and coffee; and numerous larders for different purposes – meat, fish, dry goods and often a salting room for preserving meat and a game larder, the latter often outside the main house (*see* Chapter 6). The kitchen area was under the charge of the housekeeper, who had her own sitting room, as did the butler. His domain, apart from his own pantry, included rooms for cleaning cutlery, brushing clothes and storing the silver, as well as extensive wine cellars. The main bell boards, whether for sprung or electric bells, were often placed adjacent to his pantry, and by the end of the 19th century he also often had charge of the telephone room as well (*see* Chapter 7). Common ground for the servants was the servants' hall, where meals were served, the upper servants often taking their dessert course to their own private rooms. In addition, there was usually a bakehouse, brewhouse and dairy, often separate from the service wing, together with coal stores, wood stores and a lamp room for the hall-boy or footmen to trim candles and fill oil lamps (*see* Chapter 4). Servants had to work out how to use this multiplicity of rooms as effi-

ciently as possible; Peter Brears has attempted to chart the possible routes in different layouts of service areas, while the many plans redrawn by Jill Franklin give some idea of the size and complexity of service areas in the 19th century.[15]

The separation of this large body of servants from the family played an increasingly important role in country house planning as the 19th century progressed. Kerr's maxim, 'The family constitute one community; the servants another. Whatever may be their mutual regard as dwellers under one roof, each class is entitled to shut its doors upon the other, and be alone,' is well-known.[16] His insistence that the family should have free passage without encountering the servants and that the servants should have access to all their duties without coming unexpectedly upon the family and visitors was mirrored in the plans of many service areas, and other structures were often added, such as separate service staircases, to keep the servants invisible. Such a development can be seen at Lanhydrock, in Cornwall, where a fire in 1881 forced Lord Robartes to rebuild much of the house and he took the opportunity to add a new service wing, which could have been planned with Robert Kerr's 1864 book in mind. The stone staircase for men linked the lamp room, luggage lift landings, sluice room and male servants' bedrooms, while the women's wooden staircase connected the kitchens and nurseries with the female bedrooms (Fig 1.4). Both staircases connected their respective male and female

Fig 1.4
The service staircases at Lanhydrock House. The male staircase (right) is constructed of stone, since luggage was carried up that way, while the female staircase on the left is of wood.

Fig 1.5
A group of servants holding the tools of their trade outside the west front at Erddig in 1912. The owner, Philip Yorke II, is sitting at the window behind them, along with his wife Louisa and their sons, Simon IV and Philip III. The Yorkes never distanced themselves from their servants, having their portraits painted and writing poems about them.

areas to the servants' hall – the only place where staff could meet on common ground.[17] The floor plans in the guidebook to the house point out this careful attention to gender separation.[18]

The house steward, and later the butler and the housekeeper, ruled over their gendered domains. At Brodsworth Hall, in Yorkshire, the female servants' bedroom corridor was curtained off at one end to prevent their being seen from the stairs; the curtain pole still survives, and the housekeeper's room was strategically situated so that she could keep an eye on her female staff. In Chirk Castle, Wrexham, modernised in the second half of the 19th century, all the maids' bedrooms were on the attic floor and again supervised by the housekeeper. Attention was also given to separation below stairs: at Chirk, a special door was created off the kitchen which could be closed to ensure that the men making deliveries of meat to the meat larder were not seen by the kitchen maids. Similar arrangements were made at Manderston, in the Scottish Borders, built in the 1890s with many advanced technological features, one of which was utilised to prevent the scullery maid coming into contact with tradesmen delivering goods. She could open the back door using a lever just inside the scullery door and from the safety of her domain could direct the man into whichever of the various larders was appropriate.

The 'baize door' separating the family and servants is very obvious in many houses, for example at Chirk Castle, where the decoration and furnishings change dramatically from one side to the other. One exception is Erddig, Wrexham, where, as Merlin Waterson has pointed out, little alteration was made to the arrangement of domestic quarters between the 18th and 19th centuries and only a short flight of stairs separated the family quarters from the servants' areas, while the servants' hall looked out onto the main entrance courtyard. Waterson's suggestion that there might have been differing regional traditions of domestic service in the West Midlands and border country may be the case, but the difference was more probably due to the nature of the Yorke household, with its propensity for writing poems about, photographing and doing paintings of their servants (Fig 1.5).[19] Even Kerr acknowledged, albeit reluctantly, that occasionally an architect was forced to bow to the 'peculiarities' of his employers and adapt his preferred plans.[20]

In many houses, though, the efforts made to keep the servants out of sight of the household

involved the construction of service tunnels which are perhaps the most obvious sign of adherence to the principle of privacy. Several houses in Northern Ireland exemplify this. At the remarkable Gothic-cum-classical house of Castle Ward, in County Down, no servants lived in the house; they all slept either in rooms near the stable yard or came in from the village, apart from nursery maids, who slept, with the children, on the second floor (Fig 1.6). A tunnel led from the stable yard complex to the servants' entrance at basement level, which also had a number of vaulted store rooms off to each side, including two used for coal storage. At Myra Castle, also in County Down, a tunnel top-lit by ships' portholes linked the two-storey servants' wing to the model farm (*see* Chapter 2). Further west, at Castle Coole, built in County Fermanagh for the 1st Earl Belmore by James Wyatt, the basement service area was linked to the stable yard some distance from the house by a tunnel broad enough for a horse and cart. As at Castle Ward, the tunnel had vaulted chambers each side which were used as store rooms (Fig 1.7, *see also* Fig 8.1).

Undoubtedly, though, the greatest tunnel-builder of all was the eccentric 5th Duke of Portland. Living in carefully maintained seclusion at Welbeck Abbey, Nottinghamshire, he began a huge building programme in the 1860s,

comprising a set of underground rooms including a vast ballroom or picture gallery, lit by 32 gasoliers, and a very long statue gallery and library which also incorporated glazed roof lights.[21] He also constructed 4km of underground tunnels, the largest of which had grand

portals and was wide enough for two horse-drawn carriages to pass side by side; this was intended to enable him to reach the boundaries of his estate, towards Worksop, and seems to have been intended more for his own privacy than keeping his servants out of sight.[22] These tunnels were lit both by circular, plate-glass skylights and gas lamps fed from his large gas-works (*see* Chapter 4). Many other houses contain service tunnels, including Lyme Park in Cheshire and Calke Abbey in Derbyshire, where there was even a tunnel for the gardeners to reach the kitchen gardens without being seen from the house.

This is all a far cry from the medieval hall where the lord of the household kept control by visual contact with his household on a daily basis. Such contact was minimised – and, indeed, eliminated as far as possible except for senior staff – in the 19th century by careful house planning, but this could only really be achieved with the help of new technology such as better heating, which made all the rooms more useable, as well as better communications, which are further considered later in this book.

Changes in the composition of the landowning class

Despite the many social changes brought about by increasing trade and industrial growth in Britain, the ownership of a country house in its own estate was still a desirable status symbol even towards the end of the 19th century. Land, though, no longer necessarily conferred power as it had done in earlier times, although it still undoubtedly conveyed prestige. Whilst industrialists such as William Armstrong of Tyneside or bankers such as the Rothschilds enjoyed power in their own spheres of activity, they still wanted a country house as a symbol of their improved social status. It has frequently been pointed out that, on the whole, the landed gentry and aristocracy of England had long accepted new members into their ranks so long as they obeyed the rules which governed that level of society. As Richard Wilson and Alan Mackley have put it:

So long as newcomers lived by the unwritten rules of the old landed gentry, enjoyed hunting, shooting and, before the 1770s, cock-fighting, were not extreme in politics or religion, so long as they were hospitable, appeared at the local assemblies and race meetings,

treated their tenants generously and fairly, educated their children like those of other landowners, generally showed the good manners of polite society, they were soon accepted in the county.[23]

However, certainly by the 19th century, the new aspirants to landed status could rarely buy enough land to set themselves up as large landed proprietors and enable them to enter the ranks of those listed in Bateman's *Great Landowners*, for which a minimum of 2,000 acres was necessary for inclusion.[24] Many did not wish to do so, seeking instead a house in the country with a reasonable estate of several hundred acres around it where they could indulge in a luxurious lifestyle, entertaining their friends to the hitherto more aristocratic pursuits of hunting and fishing. Mark Girouard analysed the building dates of 500 country houses from 1835 to 1889, and Jill Franklin supplemented this with her survey of 380 country houses selected mainly on architectural grounds, extending the date to 1914.[25] They both observed similar trends, with the number of houses built by new families (those who had not owned a country house for at least two generations) starting well below that for the old families, passing them in 1860–4 and pulling away from them after that. The agricultural depression of the late 1870s hit those reliant on their landed incomes hardest, and after then 'old' families accounted for less than one-quarter of the building activity. Franklin also showed that from 1895 to 1914, the builders of two-thirds of the new houses in her survey possessed fewer than 150 acres of land, and concluded, 'The survey illustrates the decline of the country seat from 1895 to 1914 and the triumph of the large house in the country.'[26]

This did not, of course, preclude 'old' as well as 'new' families adding substantially to their existing houses, something that had always happened but was accelerated in the late 18th and 19th centuries by social and technological innovations that required new structures, such as servants' wings, bachelor wings, billiard rooms, gentlemen's cloakrooms, as well as, for example, gasworks and buildings for electricity generation. As will be discussed later in this chapter, some of the old aristocratic families, such as the Dukes of Devonshire, Northumberland, Westminster and Portland, were at the forefront of such developments – technological innovations were by no means the prerogative of those who had recently acquired their houses in the country.

Motivations for change

What made some landowners adopt new domestic technology and others fail to do so? Girouard assumed, 'The pioneers were radicals, utilitarian factory owners and reformers rather than virtuoso gentlemen.'[27] As one would expect, many innovations were indeed introduced by the new country house owners, particularly those who had had experience of technological change in their professional lives. The most obvious example is Sir William Armstrong, who sought relaxation from his enormous engineering and armaments factory on Tyneside and wanted a fishing lodge on the River Coquet, which grew over the years to become the enormous mansion of Cragside, Northumberland (Fig 1.8). He was essentially fascinated by scientific enquiry but

had recognised that it had to be directed towards practical purposes. Armstrong put his interest in hydraulics to good use at Cragside, as well as trying out the new electric lamps made by his friend and fellow Newcastle man, Joseph Swan. His was probably the first large house to benefit from incandescent electric bulbs, but he did not keep his ideas to himself. Armstrong's innovations appear to have influenced the owners of other country houses in Northumberland, such as Callaly Castle and probably Alnwick Castle, to undertake electricity generation, and he exchanged correspondence with the Marquess of Salisbury at Hatfield House, Hertfordshire, for the same reason (*see* Chapter 4). Another industrialist who built an out-of-town retreat, in this case fairly close to his place of work in Wolverhampton, was Theodore Mander, whose

Fig 1.8
The entrance front at Cragside, enlarged by Norman Shaw from the 1860s to 1880s for Sir William Armstrong.

family made a fortune in the manufacture of paint, varnish and printing ink. His house, Wightwick Manor, built 1887–8 in the Arts and Crafts style, was among the first to be equipped with electricity from the outset, making use of a steam-powered generator (*see* Chapter 4).

Francis Wright, who built Osmaston Manor near Derby (Fig 1.9), was the son of a Nottingham banker and a director of the important Butterley Company, which was involved in ironworks and engineering, as well as being a director of the Midland Railway. His manor was a lavish construction dating from 1846–9, with a railway in the cellars that carried coal to a hydraulic lift, which raised fuel throughout the four storeys of the house. Wright had experience of such developments in his industrial concerns and had presumably advised his Derby-based architect (*see* Chapter 8).[28] Equally, the partners in the Worcester glove-making firm known by 1855 as Dent, Allcroft & Co prospered so well that both families were able to invest in large country estates on the proceeds. The Dents renovated Sudeley Castle in Gloucestershire, while John Derby Allcroft bought the medieval Stokesay Castle in Shropshire, but then decided to build a new house nearby. Work on this house – Stokesay Court (Fig 1.10) – finally commenced in 1889 and was completed in 1892, with ladies' and gentlemen's rooms in separate wings and a very elaborate service area, together with electric lighting, hot and cold water and heating.

Commercial enterprise also provided the funds for the building of new country houses which were technologically innovative. Foremost among these were the houses of the English branch of the banking family of Rothschild. During the second half of the 19th century, various members of the family chose to build a series of houses, all in Buckinghamshire (except Tring Park Mansion, in Hertfordshire), since from here they had easy access by rail to their

Fig 1.9
Osmaston Manor, Derbyshire, built by Henry Isaac Stevens in the late 1840s for Francis Wright, a director of the Butterley Company. The house was demolished in 1965.

Fig 1.10
Stokesay Court, built 1889–92 for John Derby Allcroft MP, a partner in the Dent, Allcroft & Co glove-making company.

London banking headquarters. The three sons of Nathan Mayer Rothschild (1777–1836), Lionel, Anthony and Mayer, built or rebuilt Tring Park, Mentmore Towers and Aston Clinton respectively, while Lionel's sons were responsible for Halton House (Alfred, 1848–98) and Ascott House (Leopold, 1845–1917), and his daughter Evelina married her cousin Ferdinand (1839–98), who was the builder of Waddesdon Manor. All these houses were luxurious retreats from the banking world and were rapidly equipped with gas, water, electricity and lifts, as will be seen in various chapters in this book.

Another banker and, like Lionel Rothschild, a fighter for the full emancipation of the Jews in England, was Sir David Salomons, later an MP and Lord Mayor of London. He entered the ranks of country house builders with the purchase of a cottage at Broomhill in Kent, which was extended first by Decimus Burton in 1829 and drastically rebuilt on several occasions thereafter, mainly in the 1850s. It was inherited in 1873 by his nephew and heir Sir David L Salomons, who was keen on scientific enquiry; as well as being a pioneer of the motor car, he was probably the first person in Britain to light his workshop with arc lamps, followed by the installation of a steam-

powered electric plant by 1884 (*see* Chapter 4). Peter Thellusson also came from a European banking family but made a fortune importing tobacco and sugar from the West Indies. He purchased the Brodsworth estate in 1790, but his complex will meant that it was his great-grandson, Charles Sabine Thellusson, who inherited the fortune that enabled him to build the Italianate Brodsworth Hall (Fig 1.11) between 1861 and 1863, equipping it with a comprehensive set of sprung bells from the outset.

In Scotland, a fortune made in trading hemp and herrings with Russia enabled William Miller to remodel the Georgian house of Manderston. His second son, James Miller, inherited the house and married the Hon Eveline Curzon, daughter of Lord Scarsdale of Kedleston Hall, in 1890. A wealthy man who doubtless wished to impress his new father-in-law, James made further additions to the house and estate, including the new stable block and Buxley Home Farm. He employed the Edinburgh architect John Kinross, who was called back again after James returned in 1901 from fighting in the Boer War to create a new north front and a bachelors' wing to house guests. This house had a vast service area and one of the largest collections of sprung bells in Britain, and it was equipped with electricity and telephones on James' return from South Africa.

Another family who made a fortune from trade were the Gibbs, originally woollen cloth merchants from Exeter, who made use of their Spanish connections to engage in the importation of guano for fertiliser from South America. William Gibbs (1790–1875) travelled frequently on business from the port of Bristol and so the purchase of the nearby Tyntes estate, which he renamed Tyntesfield (Fig 1.12), was convenient for him. William and his wife Matilda lived in the original house for almost 20 years before having it completely rebuilt between 1862 and 1864 by local architect John Norton in the fashionable Gothic style. This had gas from the beginning, together with a complex water supply, fire hydrants and a large set of sprung bells. His son Anthony continued his father's technological innovations with improved heat-

ing, plumbing, an early hydraulic lift and an electricity supply.

Finally, a fortune made in grocery enabled Julius Drewe, founder of the Home and Colonial Stores, to take early retirement at the young age of 33 with the resources to build a country seat in which to establish his rapidly growing family. The reasons for his choice of North Dartmoor are complex, but – on the advice of Edward Hudson, the proprietor of *Country Life* – he employed Edwin Lutyens to build him the granite Castle Drogo (Fig 1.13) along the lines of the castle already remodelled for Hudson on Lindisfarne. Lutyens was somewhat daunted by the size of the project, begun in 1911, but it was drastically slimmed down towards the end of the First World War when Julius' eldest son Adrian was killed on active service and his father rather lost heart. It was not completed until 1930, only the year before Julius' death, but despite its medieval appearance, was lit by hydroelectricity

Fig 1.12
Tyntesfield, rebuilt 1862–4 for William Gibbs, who travelled frequently from the nearby port of Bristol in connection with his business of importing guano from South America.

Fig 1.13
Castle Drogo, built by Edwin Lutyens for Julius Drewe, founder of the Home and Colonial Stores. The castle was greatly reduced in size from the original grandiose plans of 1911.

of iron and concrete.[29] Similar fireproofing was incorporated by Richard Coad into the remodelling of Lanhydrock, devastated by fire on 4 April 1881 (Fig 1.14). As Lanhydrock's House and Collections Manager, Paul Holden, has said, 'Today, Lanhydrock stands more as a feat of engineering than as a freestanding piece of architecture – a consequence of applying the practical and mechanical technology mobilised by the Victorian building industry together with historically condensed and aesthetic interior decoration.'[30] Reinforced concrete ceilings, 305mm thick, were supplied by Dennett and Ingles of London and installed above the family rooms to prevent the vertical transmission of fire, and the firm of Merryweather seems to have been brought in to augment the existing water supply system to supply fire hydrants (*see* Chapter 9). Perhaps understandably, Lord Robartes refused to consider the installation of either gas or electricity at the time, although a battery-powered electric bell system was included.

from the very beginning. Electric bells, telephones, lifts, state-of-the-art Edwardian bathrooms and adequate water supply were also there from the start, although the house remained rather chilly despite a coal-fired central heating system.

Many houses were remodelled following fires, often with the addition of fire hydrants, as at Kelham, in Nottinghamshire, where George Gilbert Scott had been employed to make minor alterations to a basically 18th-century house and suddenly found himself rebuilding the entire house following its destruction by fire on 26 November 1857. The new house was designed to be fireproof, the rooms being vaulted in stone or brick, and in some cases had fireproof ceilings

Similarly, the major house of Cliveden, Buckinghamshire, was bought in 1849 by the Duke of Sutherland for his wife and burnt down during redecoration in the same year. Sutherland brought in Sir Charles Barry to rebuild it, who had already been working on his vast mansion at Trentham in Staffordshire, and various innovations were gradually incorporated, including gas and water supplies (Fig 1.15). The massive clock tower, built in 1861 by Henry Clutton and modelled on Barry's design at Trentham built 20 years earlier, held a tank holding 17,000 gallons

Fig 1.14
Engraving by George Montbard for the Illustrated London News, *16 April 1881, of the fire which had devastated Lanhydrock on 4 April 1881.*

Fig 1.15
Cliveden was largely
rebuilt by Sir Charles Barry
following the fire which
severely damaged the
house in 1849, the year
in which the Duke of
Sutherland purchased
the house for his wife.

Fig 1.16
The surviving wing of
Mount Stuart, Isle of Bute,
following a fire on
3 December 1877. The new
house alongside was rebuilt
by Sir Robert Rowand
Andersen for the 3rd
Marquess of Bute.

of water which was pumped under the Thames from an artesian well in a model farm the other side of the river (*see* Chapter 3).

On the Isle of Bute, the central block of the 3rd Marquess of Bute's Mount Stuart was destroyed by fire on 3 December 1877. The wealthy Marquess, who had employed Alfred Burgess to rebuild Cardiff Castle only five years before, now brought in Sir Robert Rowand Andersen of Edinburgh to build an equally extravagant neo-Gothic house on the same site, retaining only the north and south wings of the old house (Fig 1.16). Construction included the building of a railway from Kerrycroy a mile away on the coast to enable the sandstone and timber from the Bute workshops in Cardiff to be imported more easily. Electric light was installed, although not without problems, and one of the first large swimming baths in a country house was built in the basement.[31]

On the personal level, interestingly, bachelordom seems to have prompted some landowners to spend their time and money on the improvement of their houses. Good examples are the 6th Duke of Devonshire at Chatsworth, Derbyshire, who employed Joseph Paxton to build his great conservatory; the eccentric 5th Duke of Portland

at Welbeck Abbey, who, as seen earlier in this chapter, built underground tunnels as well as a piped water supply and gasworks; and Lord Fairhaven, who restored Anglesey Abbey, Cambridgeshire, between 1932 and his death in 1966. Equally, wives could encourage innovation.

Fig 1.17
Countess of Warwick (1861–1938), photograph by Lafayette Portrait Studios, London, 1899.

Fig 1.18
Warwick Castle's mill on the River Avon, converted to generate electricity in 1894. From Country Life, *30 May 1914.*

Fig 1.19
Audley End, a house with monastic origins which was inherited in 1762 by Sir John Griffin Griffin and rapidly modernised.

At Berrington Hall in Herefordshire, George, 7th Lord Rodney, married Corisande Guest in 1891, whose mother was one of the Spencer-Churchill family of Blenheim Palace, Oxfordshire. Lady Rodney wanted similar standards of comfort in her own home and Lord Rodney complied by installing bathrooms in a separate tower attached to the house (*see* Chapter 3). Sadly, the marriage did not last, and nor did the bathroom tower, which was removed when the National Trust took over the house in 1957. Interestingly, the addition of bathrooms at Blenheim itself had been inspired by an American heiress, Lilian Hammersley, second wife of the 8th Duke of Marlborough.[32]

In 1881 Francis Greville married the beautiful Frances Evelyn 'Daisy' Maynard (Fig 1.17), notorious for her flirtations with members of the aristocracy, including the Prince of Wales. Greville became 5th Earl of Warwick in 1893; for his wife's birthday the following year he had Warwick Castle lit with electric light driven by a turbine and dynamo in the old mill on the River Avon. This was possibly influenced by the electric lights at Chatsworth following a visit the couple had made there earlier in the year (Fig 1.18).

Another wife who convinced her husband of the need to rebuild the house with modern conveniences was Emily, wife of Edward, 10th Baron Digby of Minterne House, Dorset. She is reported to have wanted him to modernise the house immediately following their marriage in

1893, but it was the bad state of the drains by 1902 which caused Digby to commission Leonard Stokes to survey and eventually rebuild the entire house, complete with electric luggage and service lifts and one of the very first powered basement vacuum cleaners (*see* Chapter 8).[33]

Another group of lordly innovators were the newly inherited, particularly if their estates had

been neglected by their predecessors. Sir John Griffin Griffin inherited Audley End (Fig 1.19), in Essex, from his aunt, the Countess of Portsmouth, in 1762. She had done her best to deal with a sadly neglected house and had reduced it in size, but her heir – who had seen active service as a soldier fighting in Europe – set about modernising his inheritance with remarkable speed. Created Lord Braybrooke in 1788, he had already installed an early set of sprung bells, a water supply to the second floor of the house and some of the very first Bramah water closets ever made (*see* Chapter 3).[34]

At Powis Castle (Fig 1.20), in Powys, the 3rd Earl of Powis, Edward Herbert, had preferred to lead a quiet life and had done little about modernising the house, with the result that the 4th Earl, who inherited the estate in 1891, brought in G F Bodley to transform several rooms and insert large numbers of bathrooms into the medieval fabric of the castle, as well as install central heating and electricity. Similarly, the absences of the

7th and 8th Earls of Stamford from their Dunham Massey estate, Greater Manchester, from 1845 onwards led to considerable neglect, and when the 9th Earl of Stamford returned to the Hall in 1905, he began a scheme of restoration to enable himself and his family to live there. He commissioned the architect Joseph Compton Hall to make extensive repairs and alterations to the house, including the installation of electricity, a fire prevention scheme, reconstruction and refitting of the kitchens and, as at Powis Castle, fitted bathrooms made possible by the provision of hot and cold water throughout the house.[35]

In other cases, estates previously encumbered by debt were leased out to more wealthy owners, who undertook technological innovations; one such was Lord Howard de Walden, who leased Chirk Castle from the Myddletons in 1911 (Fig 1.21). As part of his lease, he undertook to install electricity throughout the house and to provide a hot water supply and bathrooms, building 14 of them before his lease expired. He

Fig 1.20
Powis Castle, where the 4th Earl of Powis employed G F Bodley in the 1890s to modernise parts of the interior.

Fig 1.21
Chirk Castle, where Pugin
had undertaken improve-
ments in the 1840s but
which underwent further
modernisation when leased
by Lord Howard de
Walden from 1911.

The old castle was to take on a new lease of life. Wands were waved over it, which in a surprisingly short time transformed it into a model of comfort and luxury. Electricity gave light to its eyes, central heating warmed its heart, giving it an added hospitality, and a profusion of bathrooms seemed to appear from nowhere, till it became once more, surely, one of the most enviable places in the British Isles.[36]

Other innovators were just that – they were interested in modern contrivances, and they were neither radicals, utilitarian factory owners nor reformers, as Girouard had suggested. An outstanding example in this category was the 3rd Marquess of Salisbury, who was one of Queen Victoria's prime ministers and a fellow of All Souls College in Oxford. A keen experimenter with electricity, he set up his own laboratory at Hatfield House and, according to one of his staff, 'would scarcely believe it was possible to get an electric shock, until he got one himself during his experiments'.[37] As early as January 1874, he lit the gardens of Hatfield with arc lamps powered by batteries for a County Ball, and by 1882 most of the house was lit by electricity from a turbine and dynamo in his own sawmill on the River Lea (*see* Chapter 4).

and his wife Margherita were hosts to brilliant house parties between the wars. As one visitor said, comparing his later visit to an earlier one when they had worn fur coats to sit down to dinner:

Fig 1.22
Felbrigg Hall, purchased
from the Windhams by
John Ketton in 1863 and
not modernised for nearly
a century afterwards.

Similarly, the 8th Duke of Marlborough was keenly interested in science and constructed an electrical laboratory at Blenheim, as well as introducing gas, electricity, central heating and an internal telephone system of his own design to the palace – projects funded by the wealth of his American wife.[38] Algernon, 4th Duke of Northumberland, who had served in the Navy before inheriting Alnwick Castle in 1847, was interested in technological innovation and appears to have introduced a very early hydraulic system for powering lifts into his property, even before Armstrong at Cragside; the proximity of Alnwick to Tyneside and the Duke's previous naval service may have acquainted him with the possibilities of hydraulic power (*see* Chapter 8).

In the context of all these innovations, many other houses remained surprisingly unmodernised well into the 20th century, sometimes because of costs but often by the choice of the owner. Felbrigg Hall (Fig 1.22) in Norfolk was bought by John Ketton, a Norwich merchant, in 1863 after the Windham family fortunes had declined drastically, and he and his descendants seem to have been reluctant to change the house in any way: electricity arrived only in 1954 and central heating was not installed in this chilly corner of Norfolk until 1967, and then only in a small flat in which Robert Ketton-Cremer

himself lived at the time. The Yorkes at Erddig, despite various technological innovations in their estate buildings, did not introduce electricity into the house, although they did install background central heating. Similarly, the Harpur-Crewes at Calke Abbey had some of the latest technology in their kitchen garden (*see* Chapter 2) but failed to modernise the house, to the benefit of the National Trust, who are able to display it as a country house in decline. The family played little part in county society and were generally reclusive, particularly Sir John and then Sir Vauncey Harpur-Crewe between 1844 and 1924, a time when innovations were taking place elsewhere.

Another recluse was one of the last of the Drydens at Canons Ashby, Northamptonshire (Fig 1.23). Known as the Antiquary, Henry Dryden preferred not to alter the house, continuing to use the four-seater earth closet in the Pebble Court until his death in 1899.[39] Even at Belvoir Castle in the early 20th century, Lady Diana Cooper, neé Manners, could recall how in her childhood there, an occupied bedroom was serviced by a variety of servants who delivered water and coal (*see* Chapter 5). The availability of servants was clearly vital to the preservation of such a way of life, something which will be further considered later in this chapter.

Fig 1.23
Canons Ashby, home in the late 19th century to the reclusive Henry Dryden, the Antiquary.

Keeping up to date with domestic technology

Some country house owners, such as Lord Armstrong, Sir David L Salomons and Lord Salisbury, were clearly personally familiar with new developments in domestic technology. How did others, however, become aware of the possibilities of heating, lighting and other domestic conveniences which they could have in their existing or newly built country houses? This brief section will consider some of the possibilities but it is a topic that requires a great deal more research than was within the scope of this book.

As Wilson and Mackley have said, 'However beguiling the prospect of creating a new house, or altering an existing one, the building of a country house was always a complex task in terms of planning, finance and management.'[40] Undoubtedly, landowners sought advice from each other about their projects, as has been seen earlier in this chapter in the case of Lord Salisbury's consultation with Lord Armstrong about the installation of electricity. When John Rous at Henham in Suffolk became dissatisfied with the quality of bricks being made on his estate when James Wyatt was rebuilding his house in the 1790s, he instructed his clerk of works to consult Lord Coke of Holkham, who had already helped another landowner in this respect.[41] Landowners must have met frequently in their social or professional lives at local assemblies, magistrates' courts, the Houses of Parliament, London clubs and so on, and one assumes these matters were discussed, although there is, of course, no direct evidence. Undoubtedly the Great Exhibition of 1851 enabled many to see the new products available, such as Flavel's closed stove or 'Kitchener' which won a gold medal (see Chapter 6).

Country house owners or their agents would have seen the advertisements for suppliers of domestic items in such publications as Walford's *County Families of the United Kingdom*, published under a variety of titles from 1860 to 1920, and so would have got to know about national and local suppliers. At Audley End, for example, Lord Braybrooke's agents, Nockolds and King, dealt with a firm from nearby Chelmsford when the installation of gas was being considered in 1868, a project that finally came to nothing, and with the London-based firm of Merryweather over the possible purchase of fire appliances in 1875.[42] Professional journals such as *The Builder*, begun in 1843, or *The English Mechanic*, published weekly from 1865 to 1926,

were probably read more by contractors than agents but did provide practical advice on technological matters; for example, Frederick Allsop wrote articles on electrical bell-fitting in 1887 in the latter, republished as *Practical Electric Bell Fitting* two years later, followed by a manual on telephones in 1892 (see Chapter 7).[43]

It is not easy to ascertain the exact role of architects and building contractors in informing their clients about the possibilities of domestic technology. Some clearly did, notably the extensive architectural practice of several generations of Wyatts in the latter part of the 18th and early 19th centuries. Both Samuel and James Wyatt designed metalwork as well as buildings, and James was responsible for the cast-iron heating stoves at Kedleston and Castle Coole and also designed an underfloor hot-air heating system at Dodington Park, Gloucestershire (see Chapter 5).[44] Sir Jeffry Wyatville, their nephew, was described in 1809 as an architect 'who has manifested much skill in converting the interior of old, ill-arranged mansions to the present, and more comfortable modes of domestic life'.[45] He was employed to do just that at the late 16th-century house of Wollaton Hall, Nottingham (Fig 1.24), building a new service wing, improving the drainage and creating additional security measures on the estate (see Chapter 9).

Books on country house planning began to include advice on technological installations after the middle of the 19th century. John Claudius Loudon, a prolific writer on house planning, garden design and horticulture, was more concerned with villas than country houses, but gave some advice on the installation of stoves, gas lighting, bells and water closets, as did Kerr. By 1880, Stevenson was well aware of technological developments and included chapters on heating, ventilation, artificial lighting, water supply, hot water service, house sewage, bells, speaking tubes and even lifts in his *House Architecture*.[46] By the early 20th century, such items were considered by architects as a matter of course. In 1907, four years after they had been called in to modernise Polesden Lacey, in Surrey, by Sir Clinton Dawkins, the architects Ambrose Poynter and Stephen Terry, of the Institution of Civil Engineers, contributed an article to *The Engineer*. In this, they pointed out:

> The requirements of comfort and hygiene demanded today in country houses far exceed anything dreamed of when many of these houses were built, over a century ago, and

when their modern inheritors or purchasers contemplate living and entertaining in them, the aid of the engineering scene of the day has to be called in to provide those needs.[47]

It is perhaps ironic that Sir Clinton died in 1905 and that further alterations were being made to Polesden Lacey by his successors Ronald and Margaret Greville when this article was actually published! Lawrence Weaver, the Architectural Editor of *Country Life*, commissioned chapters on technical items such as heating, lighting and water supply in his book *The House and Its Equipment* in 1911,[48] while Walter Cave, the Arts and Crafts architect, was President of the RIBA's Architectural and Engineering section, and expressed strong views on domestic technology in an address reported in *The Builder* in 1914, in which he dealt with ventilation, central heating and water supply systems, and was impressed with the new vacuum cleaners.[49]

Architects and contractors were doubtless able to advise their clients on what was available but it was still up to the landowners to decide what was installed or not; at Lanhydrock, as seen earlier in this chapter, Coad advised Lord Robartes on the installation of fireproofing but the latter refused to countenance gas or electricity. Christine Hiskey has suggested that Lord Coke took a direct hand in planning domestic conveniences at Holkham.[50] Matthew Boulton, too, demanded additional bathrooms and WCs when James Wyatt presented him with designs for his new Soho House, Birmingham, in 1796.[51] Architects Anthony Salvin and William Burn must have had problems with owner Gregory Gregory when they were employed at Harlaxton, Lincolnshire (Fig 1.25), between 1837 and 1845, since the new owner played a major role himself. Loudon stated: 'The building is, however, closely watched by Mr Gregory himself, who keeps himself informed of every part of the construction; and who, entering into the design and practical details of the construction so completely, may be said to have embodied himself into the edifice, and to live in every feature of it.'[52] Harlaxton had provision for hot and cold water, hot-air heating, a comprehensive set of sprung bells and a railway for the distribution of coal (*see* chapters 5 and 8).

Specialist firms were producing their own publicity material by the later 19th century, which enabled their products to circulate more widely. Drake and Gorham, who installed electric lighting into many country houses, produced brochures which not only advertised their products but included presumably carefully chosen testimonials from satisfied country house owners who had made use of them; the firm also produced a book, *Light and Power: A Treatise*

Fig 1.24
The south-west front of Wollaton Hall; Sir Jeffry Wyatville introduced modern conveniences into this late 16th-century house in the 19th century.

on the Application of Electric Current to Every Phase of Country House and Estate Requirements (1904), which seems to have been sent to potential customers and described various of their projects in glowing terms.[53] John Blake of Accrington, makers of ram pumps for water supply (see Chapter 3), produced similar brochures complete with testimonials and illustrations of their work, while Merryweather, who supplied fire protection systems, wrote a book, Fire Protection in Mansions, in 1884, which must have circulated widely (see Chapter 9).[54]

Large suppliers of kitchen fitments and other household fittings, such as Clement Jeakes, J L Benham and Joseph Bramah and Company, not only had publicity material about their products but large showrooms in London where these could be seen. It is even possible that at least the senior servants, who frequently changed households, might have been able to suggest improvements with which they had become familiar in other houses.[55]

The 'servant question'

As seen earlier in this chapter, country house plans from the late 18th century onwards included increasingly elaborate service areas, with a growing multiplicity of rooms designated for specialist tasks. Little thought was given to the possibility of any shortage of people desiring to enter domestic service. Kerr in 1864 argued that the service area had to be planned around the number of servants required in the household: 'It is manifest that the amount of accommodation must be regulated directly by the list of servants to be kept. The list being determined, the accommodation has simply to be made to correspond.'[56] There is little evidence before the final decades of the 19th century that the introduction of labour-saving devices such as gas lighting or hot water supplies around the house might impact on the number of servants required or, conversely, that an increasing shortage of servants might force more country house owners to adopt such innovations for the sake of economy. A warning note was perhaps sounded by Mrs Gaskell when, in the novel North and South, written in 1854–5, she described the dilemma of Margaret Hale who, seeking a new servant, was forced to lower her hopes and expectations every week, as she found it difficult to find 'anyone in a manufacturing town who did not prefer the better wages and greater independence of working in a mill'.[57] Stevenson in 1880, although still planning for large numbers of service rooms, gave a hint that things were changing. He suggested rather scathingly that 'convenience' in country house planning could result in an immense saving of labour: this was 'an object of great importance, not so much to prevent the servants being over-worked (the danger of this is not at present great), but that the house may be managed with fewer of them'.[58] Bannister and Herbert Fletcher, in their *English Home* of 1910, stated, 'The convenience and completeness of the domestic departments due, no doubt, to the servant problem, also form a conspicuous motif in modern house plans.'[59] In the early post-war period, Randall Phillips, Editor of *Homes and Gardens*, could write a book entitled *The Servantless House*, in which he encouraged those running their houses with minimal staff to adopt various labour-saving devices to make themselves comfortable and eliminate unnecessary work. His book was aimed at the professional classes rather than country house owners but his observations that, following the First World War, 'girls who formerly accepted the shackle of what was little better than domestic drudgery came into a new liberty' and so were reluctant to return to domestic service applied to all employers of servants.[60] How far was a decrease in the availability of servants responsible for an increase in the popularity of technological innovation?

Careful study of the plans of country houses in the second half of the 19th century, particularly in the later decades, indicates that the service areas of houses then being built became more compact, with fewer purpose-built rooms. Equally, the number of people in domestic service was declining in relation to the general

growth in population at the same time; there were more opportunities in retailing or factory work which proved more attractive to many women, to which Mrs Gaskell had drawn attention in the Manchester area back in 1855. Jill Franklin has shown that the number of women servants between 1871 and 1881 failed, for the first time, to keep pace with the growth of the population, the latter increasing by 14.4 per cent but the number of female servants by only 1.9 per cent.[61] She has also shown that the number of male servants dropped from 56,000 in 1881 to 54,000 in 1911, even before many of them volunteered in the First World War and were lost to domestic service. The introduction of gas and electricity reduced or eliminated the need to trim the wicks of candles and refill oil lamps, which had been done by male servants. Female servants were increasingly spared the drudgery of carrying hot water to bedrooms or emptying slops. But more detailed analysis of individual households has shown that many families did not in fact reduce their households before 1914, a point made strongly by Jessica Gerard, and domestic service remained the major employment for women until well into the 20th century.[62] Even at Calke Abbey, which under the reclusive Sir Vauncey Harpur-Crewe did not retain a large domestic staff, the majority of those shown in Fig 1.26 were female and the lesser domestics like the kitchen maids were probably not included in the photograph.

Fig 1.26
A group of male and female servants at Calke Abbey in the early 20th century.

An initial assumption in the research for this book was that country house owners had been cushioned from the necessity of introducing labour-saving devices by their large household staff, but that the increasing difficulty of obtaining servants later in the 19th and early 20th centuries was responsible for their change of attitude. Stevenson in 1880, far more interested than Kerr had been in modern appliances, felt that this was the case, and argued that they represented an enormous saving in labour, referring again to the perceived shortage of servants. However, he sounded a note of caution: 'Such machinery may be overdone. The more there is of it, the greater the need for intelligent management, and the chance of its going wrong.'[63] As will be seen later in this book, whereas in some cases skilled men were employed to manage gas or electricity lighting plants, in others it was left to the existing servants. Dudley Gordon, discussing refrigeration in 1911, said, 'Such machinery requires very little attention and can be looked after by a person who has other duties to perform; for instance, one plant in a certain country house had been run and looked after by the laundress for the past two years.'[64] It is difficult, in fact, to tell how far new technology was perceived mainly as a way of saving labour. Jill Franklin suggests that this was never the case but that country house owners would only install new appliances if they thought that their comfort might be increased by their doing so.[65] However, as seen in this chapter, the reasons

why technological innovation took place in some country houses but not in others was very much governed by the circumstances and interests of individual landowners. It does seem to be the case, though, that the falling incomes of the landed gentry following the agricultural depression of the 1870s, together with the alternative avenues of employment open to young girls in particular, made the employment of vast numbers of servants increasingly prohibitive in the final decades of the 19th century. The First World War exacerbated a trend already in existence and by the 1920s, labour-saving devices, often previously thought of as the eccentricities of particular individuals, now became a necessity, hence the enormous increase of advertisements for powered laundries and vacuum cleaners in journals available to the gentry.

In summary

The social, economic and moral developments of the late 18th and 19th centuries were, then, reflected in the composition of the landowning classes, the planning of the many country houses which were built or rebuilt in this period, and the motivation of their owners for the changes which took place. The remaining chapters of this book provide a more detailed survey of the technological innovations introduced into these country houses, but they must be considered within the context discussed in this chapter.

2

Beyond the house: technological innovation in estate buildings, parks and gardens

'To inspect the improvements carrying on upon each other's farms has led to a social intercourse which could scarcely have been formed under any other circumstances.'

William Pitt, 1817[1]

Despite the progress of industrialisation in Britain from the mid-18th century, the landed estate still employed a large proportion of the rural population and remained one of the foundations of the country's economic activity until at least the end of the 19th century. This chapter is concerned with the technological innovation on the estate itself, which often preceded that of the house, but it does not focus on the role of the estate within the wider economy, which has been considered elsewhere.[2]

Landowners often let the majority of their lands out to tenants, rents forming a considerable portion of their incomes, but nevertheless most of them retained a deep interest in the running of their estates. Many of them were living in houses which had been built in the Palladian style and, given the experiences of many landed gentry on the Grand Tour in the late 17th and 18th centuries, they may well have been conscious that Palladio's villas in Italy incorporated buildings for aspects of farming husbandry, such as dovecots and granaries. The majority of owners had benefited from a classical education and were familiar with the emphasis in classical writing on the importance of the country estate, even to those who had held public office in Rome or its provinces. The poet John Dryden translated Virgil's Georgics, much of which dealt with arable and animal husbandry, into English verse in 1697, which may well have sparked a renewed interest in country life among the more educated classes during the 18th century. Another translation was made in 1742 with the object of inspiring the sons of gentlemen to improve agriculture, although it is unlikely that the Georgics were viewed as an actual handbook.[3] Nevertheless, J M Robinson has suggested, 'The landowner viewed himself as a latter-day Roman, the inheritor of all the classical literary resonances pervading agricultural and rural life; he saw his house, park and farms as a recreation of Pliny's villa at Tusculum, Horace at Tivoli or Virgil's Georgics.'[4]

However, most aristocratic owners of great estates had to spend a proportion of their time away from their country houses on matters of regional or national politics, as well as pleasure, and so were obliged to leave their lands in the care of their steward. By the late 18th century, with the advent of more scientific techniques in agriculture, greater exploitation of woodland and often of mineral reserves on estates, the steward had often been replaced by a full-time, professional land agent. This was usually a surveyor with experience in the valuation of land for enclosure, tithe calculation and, later, railway land purchase.[5] The agent managed his own staff, including the heads of other departments required on the estate, such as the home farm, forestry, game preserves and both the ornamental and kitchen gardens. Decisions about the kinds of technological innovations discussed in this chapter may well have been made, ultimately, by the landowners themselves, but there would have been considerable advice and often pressure from their land agents or stewards and from the staff they controlled.

Undoubtedly, however, it was the landowners themselves who were the main driving force behind innovations in their parks and gardens, many of them, of course, inspired by what they had seen on the travels in Europe which many of them had undertaken in their youth. They wanted lakes, fountains, cascades and so on to provide a magnificent setting for their grand country houses, but relied on their architects and landscape designers as well as their agents and gardeners to achieve their desires. Schemes for the supply of water to these features were often taken more seriously than the supply of water to the household, pumps being developed for fountains and cascades at a time when servants in many country houses were still fetching the water for use on the upper floors from taps in the basement.

As will be seen in Chapter 4, in the 19th century owners of country estates often had to provide their own gas and electricity if they wanted to install technological innovations that required them, generally being too far from the nearest town to obtain supplies. The generating stations were often built, run and maintained by estate staff with external advice, continuing the long-standing tradition of a country estate's economic self-sufficiency. Produce grown on the estate had long been processed there for home consumption; this included grinding grain for animal feed and flour, malting barley, brewing beer and sawing timber. If the geology of the estate was suitable, stone could be cut and sawn and bricks and lime produced for buildings,

Fig 2.1
The Perseus and Andromeda Fountain, commissioned from William Nesfield in 1853 by Lord Dudley of Witley Court.

manufactured on the estate in wood or coal-fired kilns. The kitchen garden supplied fruit, flowers and vegetables to the house in greater amounts as households grew larger and more entertaining was done. The increasing sophistication of the owners and their wives in matters of food resulted in ways having to be found to grow exotic fruits or out-of-season vegetables, and so it is not surprising that technological innovations in heating as well as in glass and iron construction were first used in the kitchen and walled gardens rather than in the house itself.

Unlike established families, many of the 'new' owners discussed in Chapter 1 did not have a long association with the land, having achieved their status in other fields, such as industry and commerce. They did not rely extensively on income from estate rents and were generally less interested in agricultural innovation or their home farms, but were undoubtedly passionate about both their ornamental gardens and their kitchen gardens since their country seats were places for entertaining friends and clients. Some, like industrialist William Armstrong, owner of Cragside, in Northumberland, could apply their professional expertise to the improvement of their kitchen gardens. Others had the money to indulge their tastes, like Lord Dudley at Witley Court, Worcestershire, whose fortune made in heavy industry and railway construction enabled him to employ William Nesfield to transform the gardens and construct the vast Perseus and Andromeda Fountain in 1853 (Fig 2.1).

Technological innovation in the estate, park and gardens played an important role in introducing landowners to the possibilities of improved water supplies, heating systems, ventilation and new building materials for use in their houses as well. By the late 19th century, some of their estate buildings were transformed to improve domestic comfort, with corn and sawmills often converted for the generation of electricity.

Processing the products of the estate: corn mills, sawmills and dovecots

The ability of landowners to force their tenants to grind corn in manorial mills is, of course, a legacy of the feudal economy dating back at least to the 11th century.[6] Watermills were listed in the Domesday Book as part of the landowner's property; windmills came later and played a much less important role in the manorial economy. The monopoly of manorial mills began to lapse in the 15th century, but since their watermills had utilised prime water-power sites from the beginning, many remained in use for the estate and its tenants.

Prior to grinding, corn also had to be threshed, a process long carried out with hand flails until Andrew Meikle patented his mechanical threshing machine in 1786. This was often powered by a horse-gin, and such machines were occasionally housed in distinctive circular or hexagonal buildings attached to barns; these machines could also be used for processes such as chopping roots for cattle feed, and so can be found within estate buildings.[7] William Pitt, one of the group of men who compiled reports for the government on improved agriculture in the final decades of the 18th century, described in about 1805 a six-horse-power threshing machine on Lord Talbot's farm at Ingestre in Staffordshire, which also had a smaller one in a barn and a portable one for moving to where it was needed.[8] Water power and later steam power was adapted to drive these threshing machines, and a remarkable example of one driven by water power survives as part of the model farm attached to Myra Castle in County Down (Fig 2.2).

At Shugborough, Staffordshire, the home farm complex was designed by Samuel Wyatt, who seems to have played a role in devising the water power systems for both threshing and

Fig 2.2
The waterwheel driving a threshing machine at Myra Castle, Strangford Lough.

grinding corn. The River Sherbrook had previously been diverted by the Ansons to supply water to various mills on the estate, and Wyatt altered the watercourses once again to provide power for a threshing machine at White Barn Farm (excavated in the 1990s) and a corn mill, fed from a mill pond, as part of his planned Shugborough Park Farm in 1805.[9]

One example which still functions commercially as a water-powered corn mill is Charlecote Mill on the Fairfax Lucy estate in Warwickshire. A mill at Hampton Lucy was mentioned in the Domesday Book, but the present mill building and mill house were built about 1806 by the Lucy estate, and are still owned by Sir Edmund Fairfax Lucy. It has been run separately from the estate since 1978 and once again produces stone-ground flour. Stainsby Mill on the Hardwick Estate in Derbyshire is, on the other hand, run by the National Trust as part of the estate, and was substantially restored in 1850 by the 6th Duke of Devonshire, who spent a considerable amount of money on it. In Cornwall, the mill on the Cotehele estate is also in the possession of the National Trust and has a fine overshot wheel fed from the Morden Stream. The mill site has been in use since medieval times but the present building dates from the 18th century and was in regular use until 1964 for grinding corn for cattle food. As on many estates, the mill is part of a wider complex of estate buildings, like that at

Castle Ward, County Down, on the shore of Strangford Lough. The mill here was originally a tide mill dating from the early 18th century and contains a kiln for drying grain as well as conventional mill machinery. It was incorporated into the range of buildings reconstructed in the 19th century in splendidly castellated style (Fig 2.3).

The transformation of corn mills into saw-mills by many estate owners in the mid-19th century may well have been due to the increasing imports of cheap grain from the New World, which was often ground in roller mills at the ports for national distribution. It is noticeable that many country estates increased the amount of timber they grew in the course of the 19th century as grain prices fell, processing their estate crop both for domestic use and for sale. Water-powered sawmills represented a considerable advance on the old saw pits, particularly when they were equipped with frame saws which made use of an inching mechanism to move the timber forwards under the saw. A splendid example still exists at Dunham Massey, Greater Manchester, where the old corn mill was converted for sawmilling in the mid-19th century (Fig 2.4). It is the only part of the extensive building works carried out by Sir George Booth in the early 17th century to survive. The mill dam is part of the lake at the front of the house, which may at one time have been a moat round an earlier house. The mill ground corn until the

Fig 2.3
The corn mill, originally a tide mill, and other farm buildings at Castle Ward.

1860s, when it was converted into a sawmill, perhaps because another corn mill on the estate nearer to Bollington was being rebuilt on a large scale. At the same time, a pump was also placed in the mill to provide water for the house, replacing a hand pump in the gardens (*see* Chapter 3). The mill contains machinery including a frame saw, band saw, circular saw, boring machine and a lathe, which are still driven by the waterwheel.

Another mill which served several purposes was built in the late 1840s at Osmaston Manor in Derbyshire. Here, the high breast-shot iron waterwheel drove pumps to supply water to the house, as well as driving the sawmill (*see* Chapter 3 and Fig 3.5). Florence Court, not far from Enniskillen in County Fermanagh, has a splendid collection of estate buildings which include a water-powered sawmill also built in the 1840s to process the estate timber, which continued to work until the 1950s. A slightly earlier water-powered sawmill survives in the park at Gunton Hall in Norfolk, the former residence of the Harbord family, created Barons Suffield. The third Baron Suffield, who inherited in 1825, was another estate owner who increased the amount of woodland on his estate and presumably built the sawmill near the entrance to the park soon

afterwards, as it was threatened by the Captain Swing rioters in 1830. The mill ceased working in the 1950s and over the years became derelict. In 1976 it was decided to demolish the mill and the grist mill machinery it contained. Fortunately, the Norfolk Windmills Trust purchased the lease in 1979 and, in partnership with Norfolk Industrial Archaeology Society, restored the mill, which is now open to the public on several occasions throughout the year.

The water supply for many estate-based corn and sawmills found a new use in the late 19th century for the generation of electricity. One of the first electricity installations in a country house was made possible by the Marquess of Salisbury's conversion of his sawmill at Hatfield House, Hertfordshire, in 1882, powered by the River Lea; the site of his turbine is still visible. At Bateman's, East Sussex, a turbine was inserted alongside the existing waterwheel on the corn mill to provide electricity (*see* Chapter 4). John Claudius Loudon waxed lyrical about the mill at Warwick Castle, which he felt added to the picturesque landscape. He claimed, 'The water-wheel and corn-mill at Warwick castle is perhaps the grandest appendage to that noble building.'[10] This was converted for electricity

generation in 1894 (*see* Chapter 1 and Fig 1.18). The splendid corn mill at Mapledurham, part of the Mapledurham Estate in Oxfordshire and recorded in the Domesday Book, is the last working watermill on the Thames (Fig 2.5). The house has a 15th-century core, as does the mill, whose size has been increased several times in its history. One of the waterwheels continues to drive millstones for flour, but the second waterwheel was replaced during the 1920s by a turbine which produced electricity for the estate, and this has now been superseded by an Archimedean screw serving a similar purpose.

Dovecots were, like corn mills, part of the earlier manorial economy when young birds provided a welcome source of tender meat, since pigeons bred rapidly. Keeping them was the privilege of lords of the manor and monasteries during the Middle Ages, a source of considerable conflict as the birds ate the crops of lord and peasant indiscriminately, but these manorial privileges had ceased at least by the 17th century. Freestanding dovecots were generally built of stone or brick, meaning that many still survive and can be seen in parkland, walled gardens or even the estate yard, as at Dunham Massey (Fig 2.6). The arrival of the brown rat in Britain in the early 18th century, a voracious predator, resulted in nest holes being built out of their reach and often pigeon lofts were incorporated at a high level into barns and other buildings. Once agricultural innovations made it more possible to keep livestock through the winter, dovecots were really redundant and the buildings became more of an ornamental feature rather than an economic necessity.[11]

The estate yard and model farms

By the 19th century, the estate yard of large country houses had become the hub of a large-scale economic enterprise and its buildings reflected the intense activity required to run both the estate and the house itself. Workshops for the maintenance of farm machinery – a forge, wheelwright and blacksmith's shops – were needed, together with some stabling, the brew-house and often the laundry, dairy and bakehouse (*see* also chapters 3 and 6). In many cases, buildings were added as they became necessary; in others, especially in important grain-growing areas, they were carefully designed to maximise efficiency, particularly with the advent of mecha-nisation. Some were architect-designed to present a uniform appearance and have become known as model farms, such as those designed by Robert Adam at Culzean Castle, Ayrshire, and by Sir John Soane at Wimpole Hall, Cam-bridgeshire. Many model farms at the turn of the 19th century were the work of Samuel Wyatt, who combined a liking for neoclassical architec-ture with a strong interest in mechanical devices, as seen earlier in this chapter in the case of the threshing machines at Shugborough, where he also built the model home farm, complete with a tramway to carry threshed grain from the corn mill to the feed store (Fig 2.7).[12] Wyatt was also responsible for many of the farm build-ings for Thomas William Coke at Holkham, Norfolk, such as the impressive range of work-shops grouped around a central clock tower at Longlands Farm.

Robert Salmon was the architect for model farm buildings for Coke's great rival as an agricul-tural improver, the 5th Duke of Bedford at Woburn, Bedfordshire. Salmon equipped Park Farm there in 1797 with up-to-date machinery, including a mill with a waterwheel, which he opened up for inspection at the annual sheep shearings; similar events were also held at Holkham.[13] Such events spread the word about technological innovation: as Pitt said of landown-ers in Staffordshire, 'To inspect the improvements carrying on upon each other's farms has led to a social intercourse which could scarcely have been formed under any other circumstances.'[14]

Most country estates possessed their own brewhouse and often their own malt mill and malt store since they were catering for large households where a beer allowance was a nor-mal part of servants' wages. The brewhouse was

Fig 2.7
The pigsties and farm buildings in the model farm at Shugborough, designed by Samuel Wyatt.

often placed in the estate yard close to the house or formed part of the stable yard complex, an early example being that built in the stable yard at Lacock Abbey, Wiltshire, by William Shar-rington during his conversion of the Abbey following the Dissolution of the Monasteries (Fig 2.8). Robert Kerr believed, 'The Brewhouse ought to be so placed that its vapours shall not penetrate into and around the house,' while Loudon had earlier advised that it should beplaced close to the bakehouse and wash-house so that the flues from all of these could share one chimney.[15] The brewhouse at Charlecote Park, Warwickshire, for example, is in the western wing of the stable buildings and adja-cent to the laundry, both generating quantities of steam. At Calke Abbey, Derbyshire, the

Fig 2.8
The brewery originally built by William Sharrington c 1539 in the Stable Court at Lacock Abbey.

brewhouse is also in the stable yard, but connected to the main house some distance away by a substantial tunnel, enabling barrels of beer to be trundled straight down to the cellar.

Brewhouses were usually of two storeys, with interior timber staging to support the various brewing vessels. Water was pumped up to a tank at the top of the structure, from which it was fed into the mash tun, usually a large, round, open stave-built vessel in which water was mixed with malted barley to begin the fermentation process. Not many estates made their own malt as well, one exception being Shugborough, where a maltings was placed adjacent to the corn mill in Wyatt's model farm. Most estates bought in their malt but, as it was better freshly ground for the brewing process, many did have their own malt mills. That at Calke Abbey is positioned on the first floor in a room next to the brewhouse, so the malt could be taken across at this level straight to the mash tun. The liquid, now known as wort, flowed from this into a large copper where the hops were added; this had to be boiled and the copper was usually situated above a furnace fired with coke or coal. Good ventilation was obviously important in brewhouses, which can often be recognised by their louvred windows, as in the brewhouse in the stable yard at Shugborough (Fig 2.9). Many were converted for other purposes in the later 19th century when commercial breweries began to make inroads into country house provision. Estates conserved by the National Trust and private owners often display the buildings and even the equipment used in the traditional brewing process to visitors, something that can now rarely be seen elsewhere.[16]

The provision of building materials for estate purposes was an important part of the estate economy. Mechanisation in this industry was minimal before the second half of the 19th century and it was not until the widespread use of steam excavators to work the prolific clay beds of Bedfordshire and the surrounding areas, railways to distribute their products and the development of the Hoffman continuous kilns that estates tended to look beyond their own resources for materials for routine building construction. The house itself, if brick, had normally been built of rather better materials than could be produced in clamps or Scotch kilns, Holkham Hall in Norfolk being a notable exception.[17] However, these kilns rarely survive as they were not permanent buildings and most were frequently rebuilt after firing. The remains of a clamp kiln probably used to supply the bricks for the wall around Sir George Booth's newly landscaped park at Dunham Massey has been found by excavation, and there may be many more of these elsewhere to be discovered during archaeological work.[18] The brick kilns at both Blickling Hall, Norfolk, and Calke Abbey are ruinous, and while it is difficult to preserve structures like these, they were an integral part of the estate economy and at least records of them should be made.

Lime, both as a fertiliser for newly enclosed land and as the basis for mortar for building, was in great demand in the late 18th and 19th centuries, and many landowners with limestone resources beneath their estates sought to capitalise on this by providing lime commercially as well as for estate purposes. The limekilns on the Calke Abbey estate in Derbyshire are the largest collection of such structures on any National Trust estate and have been the subject of various

Fig 2.9
The interior of the
brewhouse at Shugborough.

Fig 2.10
The lime yard and other
farm buildings at Erddig,
with the chimney from the
boiler of the steam engine
which drove the sawmill in
the background.

archaeological surveys, with some conservation work now taking place.[19] Lime had to be ground and mixed with water before use for mortar, and another adjunct to the stable yard or other out-buildings in some places was the lime yard where this process took place; one can still be seen at Erddig, Wrexham, powered by the steam engine which drove the sawmill (Fig 2.10). An unusual example of the use of the waste heat from an estate limekiln to heat greenhouses took place at Kylemore Abbey in County Galway, Ireland. The owner of the estate was a Manchester business-man, Richard Henry, who made use of an apparatus patented in 1873 by John Cowan of Dromore in County Kerry. The limekiln provided lime to sweeten the soil of the estate, which had been reclaimed from drained bogland, and the waste heat from the kiln heated water in a boiler which in turn heated the extensive hothouses in the walled garden.[20] The theme of innovative heating in kitchen gardens is discussed later in this chapter, and it would be interesting to know if the waste heat from limekilns was utilised in similar ways elsewhere in the British Isles.

The estate yard was transformed when motive power other than water power began to be used in the 19th century. Steam engines now power-ed farm machinery and distinctive chimneys appeared on farm buildings, as at the home farm on the Tatton Park estate, in Cheshire. Here, a corn mill was built in 1854, powered by a steam engine; a replacement engine was installed about 1890, when it also had to drive the sawmill and

chaff cutters to process hay for animal feed. Much of this burnt down in 1938 and the machinery was then driven by electricity, but a replacement steam engine has been installed for display purposes (Fig 2.11). During William Leigh's rebuilding of his Gothic mansion on the Woodchester estate in Gloucestershire from the 1850s until its abandonment in the 1870s, a

Fig 2.11
The replacement steam
engine for the one originally
installed in the 1850s to
drive the sawmill and other
machinery at the farm on
the Tatton Park estate.

steam engine powering a circular saw bench is believed to have been used.[21] Combe Mill, on the Blenheim Estate in Oxfordshire, acted as both sawmill and workshop for the estate, and its equipment was powered by both water and steam, the waterwheel continuing to operate until the 1950s. Restored to use both types of power in the 1970s, it is regularly open to the public.[22]

One of the most magnificent estate yards must have been that of the 5th Duke of Portland at Welbeck Abbey, Nottinghamshire, whose workshops were described by a contemporary visitor in 1878 as 'great sheds in which every kind of work is done by skilled workmen, aided by the very best machinery'.[23] This machinery was driven by steam, the waste heat from the boilers passing through iron pipes in order to heat the workshops. Timber from the forests on the estate was brought down to the yard 'using one or other of the five monster traction engines which are housed in sheds by themselves'.[24] The timber yard itself was described in 1875 as containing 'machinery for sawing timber, with upright and circular saws, saw sharpening machines and arrangements for steaming and steeping timber for preservative purposes.'[25] The Duke of Portland had used traction engines in his excavations for his underground rooms and his various building works; it was reported that 'a fleet of ponderous traction engines and steam-ploughs was frequently in evidence,' giving some insight into the use of such modern equipment in country house building projects in the second half of the 19th century.[26]

The Yorkes at Erddig undertook a great deal of new planting on the Erddig estate during the 19th century and so new powered circular sawing equipment was placed in the estate buildings in about 1850, although the old saw pit remained. The new mill was powered by a steam engine which was supplemented by an oil engine from about 1900. One of the many poems the Yorkes wrote about their servants concerns Thomas Roberts, in their service as a sawyer from 1858 to 1907, who was perhaps brought in to run the new steam engine:

> Our veteran Thomas Roberts see,
> Whose face will long remembered be,
> For, with the exception of one year,
> He laboured half a century here.
> At all such jobs as we might deem
> Are best perform'd by power of Steam
> He was the Driver of the Gear,
> And prov'd an able Engineer.[27]

The forge and blacksmith's shops, often equipped with a tyring platform as can be seen at Cotehele and Florence Court, were always kept busy shoeing horses and repairing machinery as this became more complex; again, one of Philip Yorke's poems at Erddig gives some idea of the work in his blacksmith's shop at the beginning of the 20th century:

> He fits for use by various arts
> Our waggons, implements and carts,
> Cyclists who here a visit pay
> Go forth rejoicing on their way.
> Our engine and hydraulic rams,
> Our tanks and taps and water dams,
> And locks and keys and hinges too,
> By Wright become as good as new.[28]

The need for good maintenance was greatly increased by many of the technological innovations introduced into the country estate, and the staff had to adapt accordingly.

Housing the means of transport: stables, coach houses and motor houses

Most landowners kept large numbers of horses for riding, hunting, pulling carriages and so on, and great stables were built to house them, often with advanced technological features in construction, ventilation and lighting. Horses were among the most valuable assets on an estate and, as Giles Worsley has said, 'No animal has been so favoured by architects as the horse.'[29] In the 17th century, the increasing interest in horsemanship in the traditions of *haute école* also resulted in the construction of large riding houses on many estates for the indoor exercise and training of horses, as at the Cavendish family's Welbeck Abbey and Bolsover Castle; the latter is still in existence and displays of such horsemanship can be frequently be seen there.[30]

Kerr wrote in 1864: 'The horse is obviously constituted for dwelling in the free space and fresh air of the field ... but as it is necessary to confine him in an enclosed apartment, the operations of feeding, cleansing and ventilating, become matters of contrivance.'[31] However, stable staff were very concerned about their charges being subject to draughts, and in most cases windows remained the main source of fresh air, although cowls were also installed on the roof to draw air out of the buildings (*see* Chapter 5).

building was described after a visit made by the Thoroton Society of Nottingham in 1899 as:

> A magnificent erection 385 feet long, 104 feet broad and 151 feet high. The glass roof is supported by lofty columns and the place is lighted by 8,000 gas jets. A short distance away is the 'tan gallop', a glass-roofed arcade 1,256 feet long, built for the exercise of horses in inclement weather. The hothouses are on a similarly gigantic scale, and are unsurpassed for size in England.[34]

There were 20 cast-iron columns each side of the Riding School building, and four at each end. The central part of the roof was made of glass, and the two side roofs were made of pitch pine, covered with copper tiles. The gas lighting was described by Robert White as follows: 'A circle of gas jets, 3ft 6in in diameter, surround the capital

Fig 2.12
The stable yard at Saltram, which was lit by gas lights.

Fig 2.13
The Riding School, Welbeck Abbey, a massive iron structure lit with several thousand gas jets.

Lighting had always been a problem in stables, with flammable materials like hay and straw lying around. However, since so much activity in stables took place in the early mornings, with horses being taken out for hunting and so on, gas lighting was often installed in stables as soon as it became available on country house estates, although this was often not until the mid-19th century (*see* Chapter 4). The remains of gas fittings can often still be found in stables, for example at Saltram House in Devon (Fig 2.12), Dudmaston in Shropshire, and Brodsworth Hall in Yorkshire. It was perhaps in the construction of buildings themselves that most technological innovations can be seen, with extensive use of materials like glass and cast iron from the 19th century. Loose boxes and stalls were often constructed of wood framed in cast iron, and hay racks of cast iron were common. Anthony Salvin's stables at Alnwick Castle, Northumberland, had a glazed roof over a long corridor which had loose boxes leading off it.[32]

However, the most extensive use of these materials was probably in the massive Riding School and stables (Fig 2.13) built by the Duke of Portland at Welbeck Abbey in 1869 to replace the previous 17th-century one. According to local historian Robert White, writing in 1875, the stables contained 'stalls for 96 horses' and were 'fitted in the most beautiful manner, being quite resplendent with Minton's encaustic tiles, and the polish of the brass work of the doors and mangers'.[33] The dimensions of the new Riding School vary slightly in different accounts, but the

of each column, and each circle is festooned with beautiful cut glass. A 2-inch gas pipe, with jets, also surrounds the entire building, at about 16 feet from the ground, and is attached to the columns. Altogether there are 7,500 gas-lights in this riding house.'[35] The Riding School was reportedly only exceeded in size by the one in Moscow.[36]

Carriages and coaches had long been in use for special occasions, and had undergone considerable technological change themselves over time.[37] Country house owners kept large numbers of carriages, ranging in type from travelling chariots and waggonettes to much lighter broughams, gigs and phaetons by the 19th century, and many examples of these can be seen in the National Trust's Museum of Carriages at Arlington Court, Devon.[38] Buildings to house carriages had been added to many country houses since the 17th century, either as part of the stable yard complex or attached to the stables themselves.[39] Carriage houses were generally less architecturally elaborate than stables, and often must have been used for carriage repair as well as providing shelter; the carriage house at Powis Castle, Powys, for example, retains a winch and lifting beam, probably for raising a carriage body from its chassis for repair work. The tales of 19th-century travellers include many references to broken wheels and other carriage accidents! Canopies were occasionally erected outside coach houses to enable cleaning and probably repairs under some sort of shelter. Kerr argued that some form of warmth was needed in carriage houses, either from a stove in the carriage house itself or in an adjacent harness room, which had to be kept warm and dry to maintain the quality of the leather harness.[40]

By the early 20th century, a new type of building was often added to the stable yard, the so-called 'motor stable' or garage. The transition to the motor car as a new type of conveyance was often a fast one, even if horses retained their importance for riding, racing and hunting. At Manderston, in the Scottish Borders, for example, one of the last country house stables to be built in Britain was a magnificent structure in which a barrel-vaulted roof of selected teak covered teak stalls with polished brass fittings, while the harness room had a floor of polished marble. These stables were built in 1895, but when Sir James Miller returned from the Boer War in 1901 and re-engaged the architect of the stables, John Kinross, to rebuild parts of his house, a series of garages was included. A room that was to be shared, according to the sign on

the door, by 'motor driver's [sic] and valets' can still be seen in the rear courtyard of the house (Fig 2.14).

One of the first country house owners to possess a car was Sir David L Salomons at Broomhill, Kent, in 1895 (*see* Chapter 1). He also put on the first practical demonstration of motor cars at Tunbridge Wells in the same year and founded the Self-Propelled Traffic Association, again in 1895.[41] Salomons built what may be the earliest example of a domestic garage in England, which he called his 'motor stables'. His first building of *c* 1900 seems to have been fairly simple, but this was rebuilt on a larger scale by 1902 and the designs were publicised in 1906 in the popular publication *Motor Cars and Motor-Driving*.[42] Salomons' new range of five garages was heated by hot-water-fed radiators and lit by electricity, and also had electric sockets for inspection lamps (Fig 2.15). Most importantly, since Salomons and other early car owners were well aware of how much mechanical attention their new acquisitions needed, they incorporated timber-covered inspection or 'carriage' pits sunk into a basement in their motor stables.[43] The latter also housed a forge and a 'mechanician's dressing room', where the chauffeur could stay until he was required. Private garages were also equipped with a water supply, often a glass canopy under which the car could be washed and, in some cases, a petrol pump, like the one at Tyntesfield, near Bristol.

Unlike the grand stables of the previous century, garages tended to be out of public view; Salomons' were built on the side of the private theatre he had at Broomhill. The chauffeur was required to bring the car round when desired, and early telephone switchboards indicate that some of the first extensions outside the house were to the garage; examples of this can still be seen at Chirk Castle, Wrexham, and at Castle Drogo in Devon (*see* Chapter 7). In 1906, the architect Walter Cave built a rather more visible set of garages at Ewelme Down, Oxfordshire, which were described in *Car Illustrated* as 'a model of what a private garage should be'.[44] The motor stable, as it was still called on the plans, was centrally heated and had accommodation for seven cars and seven members of staff, as well as an inspection pit and a petrol store not far away, soon replaced by a petrol pump.

Many other garages were added to country houses in the first half of the 20th century, such as those built by Edwin Lutyens for Sir Julius Drewe at Castle Drogo and Herbert Johnson at Marsh Court, Hampshire, around 1920. Quite often, however, owners made do by converting their existing and now largely redundant coach houses. One such conversion was carried out by the architect Sir Joseph Compton Hall for William Grey, Earl of Stamford, who inherited Dunham Massey in 1905. The three former coach houses, dating from 1721, were not altered externally, but internally the wall was removed from between the two most northerly coach houses, hot-water heating pipes were introduced and two inspection pits were let into the new concrete floor.[45] At Weston Park, Staffordshire, one of the buildings in the stable block was converted into a garage with an inspection pit, and the glass canopy previously erected for carriage washing was retained. A similar conversion was undertaken at Tyntesfield, since the Gibbs family had a large number of motor cars, and a glazed canopy was added outside the doors. Some stables were also converted into garages since the number of horses being kept in country houses declined drastically as the motor car became the fashionable means of transport. The mechanical attention needed for motor cars also declined, and so the later garages tended to have less in the way of elaborate equipment than the early 'motor stables' such as those built by Salomons.

Fig 2.15
The 'motor stables' at Broomhill, built in 1902 for Sir David L Salomons.

Fig 2.16
The furnaces for heating
the garden wall at
Weston Park.

The kitchen garden

Fig 2.17
The heated wall at Tatton
Park, topped by the
ornamental urns through
which the smoke exited;
behind the wall is the
Fernery, designed by
Joseph Paxton c 1850.

Fruit, flowers and vegetables had long been grown on country estates, but the increase in the size of households in the 18th century made more intensive production a necessity. In addition, the Grand Tour had introduced members of the gentry to the culinary delights of exotic fruits like apricots, peaches, melons, pineapples and oranges and their newly engaged French chefs demanded produce which had not previously been generally available. Walled kitchen gardens were therefore established, often close to the house; Loudon felt that the kitchen garden 'should be as near the mansion and stable-offices, as is consistent with beauty, convenience and other arrangements'.[46] The stable was, obviously, a good source of manure for the gardens. However, the landscape gardens popular by the early 18th century, with lawns sweeping up to the house itself, left little space for a walled kitchen garden close to the house and many were constructed a considerable distance away. A south-facing situation, preferably on a slight slope, was better suited to provide warmth, particularly for growing the more exotic fruit trees against the walls. For example, large walled gardens were built some way away from their mansions for the 1st Earl of Leicester at Holkham and Lord Anson at Shugborough in the late 18th century, probably both by Samuel Wyatt although the evidence at Holkham is not absolutely certain.[47]

The first technological innovation to help ripen fruit of various kinds was what Loudon described as a 'flued' wall, more often known as a hot wall, made of brick with internal flues through which smoke and heat from a furnace would be circulated to warm the wall. Philip Miller, curator of what is now Chelsea Physic Garden, in his *The Garden Dictionary* of 1754, proclaimed that 'hot walls to promote the ripening of fruits' are 'pretty much practised in England,' and gave advice on their construction.[48] Early walls were heated by fires in small hearths set into the walls; later, furnaces were placed against the rear of the walls and sometimes slightly below ground level to increase the heat. Few of these furnaces survive, except as traces on a wall, but there is a remarkable set of

them at Weston Park (Fig 2.16). The position of the flues is often shown by ornamental pots or vases disguising the chimneys through which the smoke exited; again, few survive but a wall surmounted by these can be seen in the gardens at Tatton Park (Fig 2.17). Loudon advised that fires should not only be lit at night but also during the day in the month of September to promote ripening.[49] Hot walls had disadvantages in that the heat was not evenly distributed and the plants against the wall could be scorched in places, and the flues became choked with soot. Nevertheless, they were widely used, particularly in kitchen gardens in the north of England and in Scotland, where conditions were not generally propitious for growing fruit.

Hot walls went out of general use during the 19th century when glass for frames and greenhouses, together with better heating systems, became more generally available. Glazed frames for forcing, with small panes of glass set into wooden frames, were being used by the mid-18th century; Thomas Coke's original kitchen gardens of the 1720s had hot walls as well as 91 glazed frames for melons, pineapples and cucumbers.[50]

Orangeries, designed to protect citrus trees such as oranges and lemons from the winter frost, had been built in the 17th century, with windows made from small leaded panes and heated by means of free-standing stoves or pans of charcoal, the fumes from which could injure the plants. Early examples of orangeries in England include Somerset House, Westminster (1609), Oatlands Palace, Surrey (1639), and Wimbledon (1649); the earliest surviving example, dating from 1677, is at Ham House, Richmond.[51] The roofs were not glazed since large sheets of glass were not yet obtainable, as can be seen in the mid-18th-century Orangery at Hanbury Hall, Worcestershire (Fig 2.18). If space permitted, such buildings could be very large, like the massive Orangery still surviving at Margam Park in South Wales, built in the 1780s for Thomas Mansel Talbot; this was 99m long and lit by 27 tall, round-headed windows.[52] The building was narrow, only 9m wide, so that the light could reach the whole interior. This Orangery was heated by coal fires with chimneys set into the back wall, and from May to October the plants were taken out via the high rear entrance and placed around the fountain in the garden.

Fig 2.18
The mid-18th-century Orangery at Hanbury Hall, decorated with stone vases in the central section and pineapple finials on the corners.

The addition of greenhouses as well as orangeries to kitchen gardens was made necessary by the greater range of exotic plants being imported into Britain – unlike citrus trees, these could not be brought out of doors in the summer. The first glazed lean-to structures were built against heated walls and popularised by Philip Miller in his 1754 Dictionary. These had glazed front walls on masonry foundations, with sloping glass roofs which still, as in orangeries, had small glass panes set between timber glazing bars.[53] Glass was subject to tax for the 100 years between 1746 and 1845, which made such structures expensive, and the blown glass used often had a greenish tinge. As both wrought and cast-iron became more readily available, iron glazing bars could begin to replace wooden ones and had a narrower profile, admitting more light. Loudon himself had developed curved, wrought-iron glazing bars, as he argued that the shape admitted more light, and included designs for curved and elliptical frames for greenhouses in his *Encyclopaedia of Gardening*, the first edition of which was published in 1822.[54] In this, he disagreed with Joseph Paxton, who became head gardener to the Duke of Devonshire at Chatsworth in Derbyshire in 1826. Paxton preferred wooden frames and demonstrated this in his so-called 'Conservative Wall', named because it conserved heat, built at Chatsworth against a pre-existing hot wall in 1848 (Fig 2.19).[55] However, iron frames were extensively used; a rare

early example is the elegant Camellia House at Wollaton Hall, Nottingham, built by Sir Jeffry Wyatville in the 1820s and making use of heating equipment and cast ironwork from the local firm of J Harrison of Derby.

In the 1830s the Chance Brothers of Birmingham developed an improved version of the cylinder glass process in which the glass was still blown but could be produced in much larger sizes of panes. This enabled great expansion in the use of both forcing frames and greenhouses, and allowed the construction of such structures as Paxton's Great Conservatory at Chatsworth. Completed in 1840, this was a vast, curvilinear structure but even then Paxton preferred wood, using laminated timbers which bent more easily for his ridge-and-furrow roof, although the interior columns, which also acted as rainwater downpipes, were of cast iron (Fig 2.20).[56] However, his Crystal Palace for the Great Exhibition of 1851, built within nine months, did make use of prefabricated cast-iron frames and promoted the idea of these, especially as Paxton himself was able to dismantle the Palace and re-erect it at Sydenham in 1854. The repeal of the glass tax in 1845 boosted demand for greenhouses as well as conservatories, and many manufacturers now began to offer them for sale, including Foster and Pearson of Beeston, Nottingham; Messengers of Loughborough, Leicestershire; and Mackenzie and Moncur in Edinburgh, among many others; all offered timber as well as iron greenhouses, and many of their products can still be found on country house estates such as Luton Hoo, in Bedfordshire, and Chatsworth. Others have been restored or built as copies of the originals, as at Tatton Park and Audley End, Essex.

Large structures such as this needed heating if plants were to survive; sunlight may have sufficed in smaller structures but many early lean-to greenhouses were heated by hot walls or freestanding stoves, as we have already seen. William Strutt's cockle stove, designed for heating factories in the 1790s, was adapted to enable the use of warm air to heat greenhouses and other garden buildings (*see* Chapter 5). Remarkably, an early example of one of these survives at Calke Abbey but it is not clear what it was meant to heat since its air ducts do not appear to have heated the Orangery to the rear or either of the Peach Houses. Instead, the air duct rises upwards and contains vents into the store-room above the stove, and was possibly used for drying the immense quantities of seeds needed for the large kitchen garden at Calke.[57]

Fig 2.19
Joseph Paxton's Conservative Wall at Chatsworth, built against an existing hot wall.

In Staffordshire, the great landowners were determined to produce quantities of exotic fruits, and Pitt reported how the neighbouring lords Gower, Harrowby and Anson were all experimenting with ways of ripening them in the 1780s. Lord Anson at Shugborough was the first among them to make use of steam in a greenhouse 'in which melons and cucumbers are produced in perfection at all seasons'.[58] An import, like cockle stoves, from industrial heating systems, steam was generated in boilers and circulated through pipes to heat garden structures. The system had to be designed carefully so that the cooled water flowed back into the boiler, and pipework carrying the high-temperature steam could be dangerous (*see* Chapter 5). Loudon gave good practical advice on how to make use of steam pipes for a variety of horticultural purposes.[59] However, in the 1835 edition of his *Encyclopedia*, he extolled the virtues of hot-water heating over steam, since the water did not have to reach boiling point before it was circulated, and was less dangerous. This was being widely used by

the 1840s and a variety of boilers were manufactured, of which one of the most popular and reliable was the saddle boiler. This was a cast-iron half-cylinder, filled with water, from which hot water circulated and to which the cooler water returned via a carefully designed set of pipes. Unusually, two mid-19th-century saddle boilers have survived in situ in Calke Abbey gardens, one adjacent to the Orangery; additionally, one of the later popular Beeston 'Robin Hood' boilers can also be seen, produced by the greenhouse manufacturers Foster and Pearson and so named because of the proximity of the foundry to Sherwood Forest.[60] Calke Abbey is a valuable example of early garden technology.

Heating was used not just for orangeries, greenhouses and various forcing frames but also for the pinery-vineries which became popular from the 18th century.[61] The pineapple was a source of fascination to contemporaries, grown not just for its taste but its appearance, often featuring as a table centrepiece. When its cultivation was first introduced into Britain, pineapples

Fig 2.20
Joseph Paxton's Great Conservatory at Chatsworth in 1900; the structure was completed by 1840.

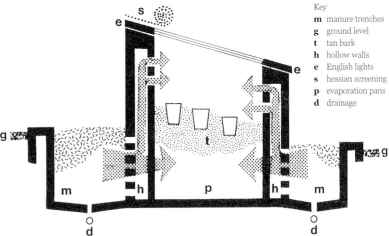

Key
m manure trenches
g ground level
t tan bark
h hollow walls
e English lights
s hessian screening
p evaporation pans
d drainage

Fig 2.21
The reconstructed Pineapple Pit, Heligan Gardens, and a drawing by James Chamberlain showing the heating system.

Fig 2.22
The hot wall of c 1861 at Dunmore, showing the vases through which the smoke exited, while the massive pineapple indicates the purpose of the building.

were forced in specially constructed 'pits' covered with glass, for the construction and working of which lengthy instructions were given by Loudon.[62] The heat was generated by waste tannery bark or manure which gave off heat as it decomposed. This was placed in trenches each side of a pit containing the pineapples, and the heat was reflected back onto them. One of these was found in the Lost Gardens of Heligan, in Cornwall, and has been reconstructed (Fig 2.21). Loudon's instructions took into account the fact that pineapples need different treatment at various stages of their growth, and the development of the greenhouse greatly aided the process of ripening. An early example is the remarkable building dating from 1761 at Dunmore in Falkirk (Fig 2.22). The rear wall was essentially a hot wall, heated by hot air through flues which exited, as at Tatton, via chimneys hidden within Grecian urns, and the topmost structure, resembling a pineapple, reflected the purpose of the building. Later, pineapples often shared a greenhouse with vines in the so-called pinery-vinery, a not altogether satisfactory arrangement as vines required more temperate conditions than pineapples. The latter were planted in heated beds within the glasshouse, while the vines were planted in a bed outside and entered the greenhouse through holes in the wall, then being trained upwards and over the underside of the roof. Good examples can be seen at Culzean Castle and Audley End (Fig 2.23), both dating from the second decade of the 19th century.[63]

It was not only the established landed gentry who took an interest in their gardens. Samuel Morton Peto, a prosperous entrepreneur, bought Somerleyton Hall, Suffolk, in 1843, reconstructed the house and engaged Joseph Paxton to build two iron-framed, lean-to greenhouses, each 27.5m long. Sir William Armstrong at Cragside became an enthusiastic gardener but the situation of Cragside meant that large areas of his kitchen garden had to be covered with glass if anything tender was to grow. An article in *The Garden Magazine*, 1872, indicates that Armstrong used his engineering abilities to achieve this:

The glass erections, which are of considerable extent, are devoted principally to the cultivation of fruit trees in pots. The engineering skill possessed by Sir William has enabled him to introduce a somewhat novel mode of treating trees in pots. In one house the floor, which is of timber, is made to rest upon wheels exactly

like a railway truck; these wheels are placed upon rails which extend sufficiently far beyond the house to allow the platform with its load to run clear out into the open air, the end of the house being so constructed as to permit this to be done. By means of the turn of a handle the whole is set in motion, and finally stands still of its own accord where it is wanted, while by reversing the handle the platform with its load moves back again to its place in the house. All this is accomplished by hydraulic pressure. The rails being laid with a slight incline to the house, the platform runs back by means of its own momentum.[64]

This device would have worked in the Orchard House, heated by a coal-fired boiler in the basement which has been removed. Heating pipes ran under raised stone platforms with holes to allow hot air to escape. On these platforms, large ceramic pots containing fruit trees stand on cast-iron discs with rollers, which allowed the pots to be rotated to ensure even ripening; these are the pots which would have been moved by the railway, but it is not certain it was successful and nothing remains today apart from the pots themselves (Fig 2.24). There were several boilers heating other glasshouses, including a cast-iron-framed Palm House, and these were all connected to a common flue which exited in a chimney across the road to the west. This flue had two intermediate hearths where fires could

be lit to increase the up-draught.[65] Armstrong welcomed visitors who came to Cragside, following in the footsteps of many of the landed gentry in the late 18th and 19th centuries who visited each other to admire the innovations in their walled gardens.

Fig 2.23
The greenhouses and vinery at Audley End, built c 1811 and now restored.

Fig 2.24
The remarkable system for turning ceramic pots on cast-iron rollers in the Orchard House at Cragside, designed by Sir William Armstrong.

The park and ornamental gardens

The chief technological interest in the many changes in garden design between the 18th and early 20th centuries was the attention given to providing sufficient supplies of water to fill large lakes and power fountains. Gardens in the 17th century demanded formal canals, ponds in geometrical shapes and classically styled fountains, whereas the landscape gardens of the 18th century favoured serpentine lakes, naturalistic cascades but fewer fountains; large fountains again became popular in the 19th century, when increased engineering skills were available to create spectacular displays. Some estates, such as Chatsworth, were fortunate in possessing a plentiful supply of water. The gardens here exemplify the various changes in landscape design over the two centuries. The 1st Duke of Devonshire engaged the garden design firm of London and Wise to reconstruct the gardens in the 1690s. They built storage ponds on the gritstone moors above the house and conducted water via leats and pipes to the formal lakes below. The well-known cascade was built in 1696 by Grillet, a pupil of Le Nôtre, who had designed the gardens at Versailles, and was enlarged six years later and Thomas Archer's Cascade House was added (Fig 2.25).[66] The fall

of water from the moors was sufficient to power several fountains in the formal gardens, including the Great Fountain in the Canal Pond. Much of the formal gardens were swept away in the 18th century and replaced by a more naturalistic landscape under the influence first of William Kent and then Capability Brown, but fortunately they spared the Cascade, Canal Pond and fountains. The 6th Duke brought in Joseph Paxton, who undertook to improve the Great Fountain in 1843 in time for a visit to Chatsworth by Czar Nicholas, Emperor of Russia (who never arrived). Paxton dug a large new lake on the moors and constructed an elevated aqueduct to increase the fall of water and a vast network of pipes to the Great Fountain, renamed the Emperor Fountain, which could reach a height of 90m until some of the outflow from it was used to power an electricity turbine in 1893 (see Chapter 4).

At Bolsover Castle, Derbyshire, a gravity-fed supply to an early fountain was boosted by a water-powered pump, the evidence for much of which can still be seen. William Cavendish, 1st Earl of Newcastle, who also owned Welbeck Abbey, transformed the gardens at Bolsover in time for a visit by Charles I and Henrietta Maria. A natural spring on the higher ground had been tapped and conducted through a pipe line with a number of small conduit ('cundy') houses along

Fig 2.25
Thomas Archer's Cascade House at Chatsworth.

its length which each contained a lead water tank. A water pipe from one of these went downhill and then up towards the castle, relying on the siphon effect to deliver water to the castle's cistern house. Here, water was pumped by means of a waterwheel to power the Venus Fountain, installed in 1628 (Fig 2.26). Similar waterwheels were used later on many country estates, including Audley End, where pumps driven by a waterwheel from the River Cam took water both to the gardens and the house by 1765. At Petworth, West Sussex, the waterwheel at Coultershaw, on the River Rother, supplied the lakes on the estate from 1782 (*see* Chapter 3). Probably the most spectacular of the surviving waterwheels is that at Painshill Park in Surrey (Fig 2.27). Charles Hamilton created the landscape garden here in the mid-18th century, installing a waterwheel to pump water from the River Mole to his lake. This was replaced in the 1830s by a cast-iron wheel built by Bramah & Sons and restored by the Painshill Park Trust in 1987. The wheel has a diameter of 10.6m, and drives pumps which still supply the 5.6ha ornamental lake.

Many books of practical information on water pumps of this kind were available, one of the most influential being Stephen Switzer's two-volume *An Introduction to a General System of Hydrostaticks and Hydraulicks, Philosophical and Practical*, published in 1729.[67] He had worked for the influential firm of garden designers London and Wise, including during the period of their commission at Blenheim Palace, where he was able to observe the introduction of Robert Aldersea's first pumping engine of *c* 1709, which was intended to provide water for fountains in the formal gardens to the rear of the house (*see* Chapter 3). Switzer went on to design many gardens, including cascades and lakes for the Earl of Gainsborough at Exton Park, Rutland, and for John Aislabie at Studley Royal, Yorkshire. Although most of the water-raising systems he described were powered by water, he became interested in Thomas Savery's 'engine for raising water by fire' and closed his account with a description of Thomas Newcomen's improvements to Savery's engine, which he thought would be beneficial for securing water for garden features.[68]

Steam engines were certainly used in the 19th century to power the water supply for fountains, one of the best-known being the Perseus and Andromeda Fountain at Witley Court, designed in 1853 by William Nesfield (*see* Fig 2.1). As elsewhere, the water supply was gravity-fed from the hills west of the house and was dammed in places to create a number of pools. The engineering was undertaken by the firm of Easton, later Easton and Anderson of Erith in Kent, who undertook many commissions for country estates, including Audley End. At Witley Court, water was pumped from the Hundred Pool by a 40hp steam engine to a reservoir in the Deer Park, from where the fountain was supplied, with a head of over 30m. Even this, however, had to be supplemented from other sources on the estate. Nesfield and Easton and Company also installed the massive George and Dragon Fountain at Holkham Hall in the 1850s, again making use of a steam-driven pump.[69]

Fig 2.26
The early 17th-century Venus Fountain at Bolsover Castle, fed by water from a spring pumped to it from the nearby cistern house.

In conclusion

Technological innovation, then, was often first introduced into the country house estate and its gardens before the benefits were transferred to the house itself. Water supplies intended initially for lakes and fountains were modified to serve the house; various types of heating systems were introduced to enable exotic fruits to be grown in kitchen gardens before central heating in the house was even considered. Machinery driven by water power or steam enabled grain to be processed for domestic consumption or timber to be sawn for use in building construction. The country house estate was seen as an economic entity which was largely self-sufficient, and so it is not surprising that when innovations such as gas and electricity came along, country house owners were prepared to apply the same principle to their production as well.

Fig 2.27
The cast-iron waterwheel, built by Bramah & Sons in the 1830s to replace the original horse-driven wheel which pumped water from the River Mole to the lakes at Painshill Park. This image was taken by Charles Harvey Combe of Cobham Park, Surrey, in 1899.

3

Water supply and sanitation

'Good water is the most simple, common and necessary aliment of animal creation, and especially of mankind, the means of procuring it have ever occasioned a proportionate degree of solicitude.'

William Matthews, 1835[1]

So fundamental was an adequate supply of water that it has dictated the location of human settlement throughout history, with towns, villages and individual dwellings clustered alongside rivers or streams, or around other water sources such as springs or shallow wells. In the medieval period, monasteries concentrated large numbers of people in one place, which probably encouraged them to develop early systems not just of water supply but also of water-flushed sanitation and sewage disposal. Castles – which also had large garrisons and often private households as well – had privies on various floors which discharged into exterior moats. The private houses which were constructed from the remains of monastic dwellings after the Dissolution often inherited these systems and continued to make use of them, for example at Audley End, Essex, and Lacock Abbey, Wiltshire. Most substantial houses dating from this period incorporated wells from which water could be drawn up by human or animal power, the remains of many of which survive, as will be seen later in this chapter. The fundamental importance of water supply meant that improvements continued to be made, particularly when house owners began to embellish their properties with ornamental lakes and fountains. Even so, in most cases, water was not available above the basements or ground floors of many houses before the 19th century, and the household relied on the work of a large number of servants to carry water around the house, as well as to remove waste. This servant army enabled ladies to continue using their private facilities of close stools and hip baths in front of blazing fires for some years after technological improvements meant that more modern facilities were available. The history of the development of water supply and sanitation in country houses is therefore a mixture of progression on many fronts accompanied by the technological inertia occasioned by the continued availability of large numbers of servants until the early 20th century.

Water supplies to country houses: pumps and tanks

As country houses grew in size and wealthier house owners sought greater privacy in more isolated locations, the existence of nearby water supplies became a less important criterion and so means had to be found for transporting water from greater distances. One constraint on this was the availability of pipes. Lead was too expensive for anything more than short lengths, so most water pipes were made of wood (Fig 3.1) – usually elm – which had a limited life; cast-iron

Fig 3.1
A section of wooden water pipe recovered from the grounds of Lyme Park.

pipes did not become freely available until the start of the 19th century. Some medieval monastic buildings are believed to have had water supplied from cisterns, fed by gravity and conducted through lead pipes, as did Westminster Palace from the mid-13th century, but this was unusual.[2] If water was to be conveyed long distances, this was often accomplished using open channels, a practice which dates back at least to Roman times; London's New River, constructed in the early 17th century to take water the 32km from Hertfordshire to the centre of the city, was a notable example. In some cases, underground tunnels were dug to bring water from distant springs. Coleshill House, on the Wiltshire/ Oxfordshire border and destroyed by fire in 1952, had a brick-lined 'water mine', built in the 1740s, which ran for 400m to an underground reservoir near the house, from where a horse-powered pump supplied water for use in the house and gardens.[3] Robert Smythson's 16th-century reconstruction of Wollaton Hall, Nottingham (see Fig 1.24), for Sir Francis Willoughby, incorporated both water supply and drainage systems, despite the Hall's somewhat inconvenient position on a hilltop. A tunnel brought water from a spring close to the site of the original house to beneath the new Elizabethan mansion, where it was raised by hand from a well in the basement.[4] Many houses made use of springs at a considerable distance to provide water for their lakes and fountains, as was seen in Chapter 2.

From the late medieval period, simple hand-powered lift pumps (Fig 3.2) provided an easier means of raising water from wells than a rope and bucket, and in humbler dwellings these sometimes provided the only supply of water until the 20th century. Wealthy households generally aspired to something better, particularly by the 18th century. Unless the house was located in a hilly area, where water could be fed by gravity from an upland source, this usually meant pumping the water to a storage tank or reservoir in an elevated location, which required more sophisticated and powerful pumps. Occasionally, these were powered by gangs of men, as was the case at Dunham Massey, Greater Manchester, but more usually they used water or animal power. Waterwheels had been used to drive pumps for mine drainage in the early 16th century but it was not until the late 17th century that they were used for water supply. In 1692 George Sorocold built a 'water engine', powered by a waterwheel on the River Derwent in his home town of Derby, supplying water around the town via a network of wooden pipes. He went on to install similar pumps in other locations, including one with four banks of four pumps, built into London Bridge in 1701.[5]

Around 1706, John Vanbrugh installed a machine very similar in design to Sorocold's, built by Robert Aldersea of London, into one of the arches of a bridge over the river running through the grounds of Blenheim Palace, Oxfordshire (Fig 3.3). This had two banks of three pumps and was originally intended to provide water for ornamental fountains in the gardens but was found to be inadequate for this purpose, so the water was directed instead to a tank in the Kitchen Court Tower, from where it supplied parts of the house.[6] As with many other types of domestic technology, this innovation was frequently utilised in the gardens of country houses before it was adopted within the houses

Fig 3.2
Manual water pump in the basement, Audley End.

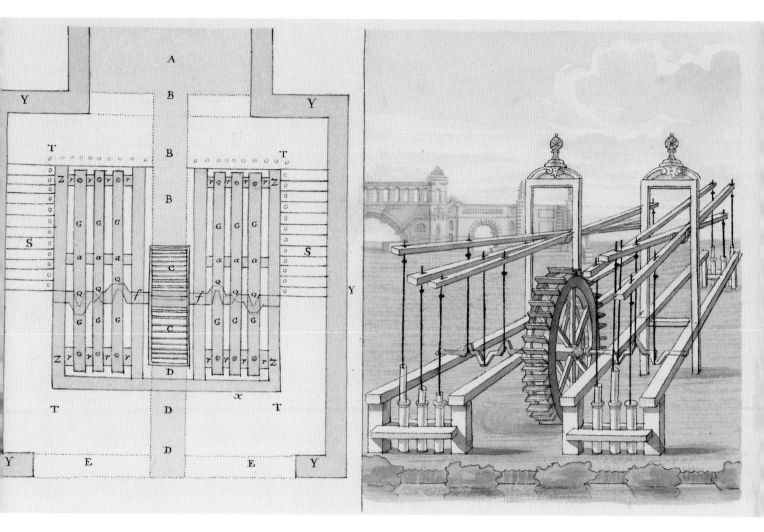

themselves (*see* Chapter 2). Another early example of this type of pump was noted by Celia Fiennes at Broadlands, Hampshire, around 1696, where there was 'a water house that by a wheele casts up the water out of the river just by and fills the pipes to serve all the house and to fill a basin designed in the middle of the garden with a spout in the middle'.[7] Petworth House, West Sussex, received water from the River Rother at Coultershaw, almost 2.5km south of the house, via pumps powered by a waterwheel, erected in 1782. This certainly supplied water to features in the gardens but may also have fed the house and parts of the town. The pump house has been restored and the pumps are operated periodically.[8]

The familiar technology and minimal running costs of the waterwheel ensured that even into the 20th century it remained a popular source of power for pumping water. At Audley End, pumps driven by a waterwheel provided water from the River Cam to the gardens and to tanks in the roof of the house from 1765; the pump house was rebuilt around 1847 and supplemented with an oil engine in 1908 but the waterwheel continued in use until the house was connected to the mains supply in 1954.[9] Interestingly, in this installation the water-powered pumps were designed to be run more or less continuously, so the supply pipe to the house was kept pressurised to feed fire hydrants around the perimeter. At Belton House, Lincolnshire, a very elegant pair of twin-barrel pumps, constructed in brass and gunmetal and driven by a waterwheel, was supplied by the London engineer John Braithwaite in 1817 to a specification by Jeffry Wyatville; the cost, not including the pump house itself, was over £1,000 (Fig 3.4).[10] One set of pumps fed spring water to the mansion and the other set supplied river water for garden fountains and fire-fighting. The pump house and waterwheel survive in situ but the pumps are in the care of the Science Museum in London.

Fig 3.3
A water-powered pump, installed c 1706 in one of the arches of Vanbrugh's Bridge, Blenheim Palace.

Osmaston Manor in Derbyshire, a very technologically advanced house built 1846–9 for Francis Wright, one of the directors of the Butterley Company, also had a dual water supply. The remarkable water-powered pump house was built at the same time as the house in the style of a Swiss chalet; the pump house survives but, sadly, the house was demolished in 1965 (Fig 3.5; *see* also Fig 1.9). As well as a sawmill, the high breast-shot iron waterwheel drove two cast-iron water pumps: one pumped water from a spring to a small reservoir on a hill above the house for domestic use, the other lifted water from the river to a larger reservoir which fed the hydraulic lifts (*see* Chapter 8) and probably also supplied garden features.

If there was no convenient stream to drive a waterwheel, animal power was often used. At Greys Court in Oxfordshire the pump house, believed to date back to Tudor times, contains an unusual vertical wooden treadmill which was operated by a donkey (Fig 3.6). Similar vertical treadwheels can still also be found at Carisbrooke Castle on the Isle of Wight and at Burton Agnes Hall in Yorkshire, both of a similar date to that of Greys Court. More common were horizontal horse- or donkey-powered 'gins', a technology frequently employed at mines for winding and pumping; the remains of one survive at The Argory, County Armagh. A common arrangement was to construct a tall pump house directly above a well and pump water directly

Fig 3.4
Twin-barrel water-powered pumps, installed at Belton House in 1817. The waterwheel survives in situ but the pumps are now in the care of the Science Museum.

Fig 3.5
Water-powered Pump House, Osmaston Manor. The waterwheel drove two water pumps and a sawmill.

into a tank in the upper part of the building, from where it could supply the ground floor of the mansion by gravity. The pump in the lower portion of the octagonal Engine House at Holkham Hall, Norfolk, is thought to have contained a donkey-powered gin; the upper tank area was later converted into the game larder. A feature in the gardens at Dunham Massey is the Well House, dating from the early 18th century, in which a pair of manually operated pumps are located on the

upper floor, along with the water tank. This was superseded in the mid-19th century, probably by water-powered pumps located in the sawmill (*see* Chapter 2).

The ram pump provided a simpler and cheaper means of harnessing water power for pumping water than waterwheels. This device is thought to have been invented by the Derby clockmaker John Whitehurst some time before 1772 and was first used at Oulton Park, Cheshire, to

Fig 3.6
A 3D CAD drawing of the vertical treadwheel at Greys Court, formerly operated by a donkey and believed to date from the Tudor period.

51

Fig 3.7
Ram pump at Ilam Hall,
Derbyshire. From Blakes
Hydram brochure, c 1921.

supply water features in the gardens.[11] White-hurst presented his invention in a paper to the Royal Society in 1775 and it was further refined by other inventors, including the Montgolfier brothers in France. Ram pumps utilise the kinetic energy of a large flow of water at a low head of pressure to operate repeatedly a check valve; the resultant pressure wave drives a small volume of water to a high head. By the mid-19th century, many manufacturers were offering these devices, one of the most notable being

Fig 3.8
The 1730s Water House,
Houghton Hall.

John Blake of Accrington. Later refinements included examples with twin chambers, which utilised river water for power but pumped spring water to the house. These pumps could operate around the clock for decades without needing any attention and many survive in situ, including the one at Ilam Hall, Derbyshire (Fig 3.7), and one which has recently been restored at Florence Court, County Fermanagh; a few even remain in use to this day.

These pumps all lifted water into some form of storage tank or reservoir in an elevated location, from where it could supply the house by gravity. As we have seen, a few of these tanks were located in the top of the pump house but more usually they were some distance from the pump. A simple solution was to place the tanks in the roof space of the house, as at Audley End, but purpose-designed water towers could also provide striking architectural features, either in the grounds or as part of the house itself. One of the most notable examples of the former is the Water House at Houghton Hall, Norfolk, dating from around 1730, believed to have been designed by Lord Herbert, later Earl of Pembroke (Fig 3.8).[12] The tower stands on a slight hill, 600m north of the mansion, and water was pumped into it once a week by a set of horse-powered pumps from a well a further 500m to the north.

Among the many houses which incorporate imposing water towers as part of their structure are Hestercombe, in Somerset, and Cliveden, in Buckinghamshire (Fig 3.9). By the late 19th century a more utilitarian approach was commonly adopted, in which water was pumped up to covered brick tanks located on a hillside, sometimes several kilometres away, from where it could be distributed not just to the house but to other parts of the estate as well; examples include Manderston in the Scottish Borders, Stokesay Court in Shropshire and Tyntesfield, near Bristol. Water tanks were sometimes fitted with water-level gauges which could be read remotely, so staff could easily see when they needed replenishing without having to climb up to the tanks; examples survive at Wallington in Northumberland and Ickworth in Suffolk, where there are gauges in the basement service areas showing the level of water in the roof tanks; and at Polesden Lacey, in Surrey, where there is a large gauge on the outside of the water tower. Manderston had an electrical telemetry system, which monitored the water level in the distant tanks via a recorder located in the Estate Office; the recorder, made by Gent of Leicester, probably dates from around the second decade of the 20th century.[13]

By the end of the 19th century, other forms of prime movers also became available to drive water pumps. Domestic water at Cragside, Northumberland, was collected from a number of springs in the hills above the mansion and fed to a pump house in the valley below. This contained a pair of hydraulically powered double-acting pumps, which pumped the water back up to a covered tank on the hill behind the mansion, from where water was supplied by gravity. The system dates from the original construction of the house around 1865 and can still be operated.

Steam power, in the form of Thomas Savery's engine, developed primarily for mine drainage, was apparently employed to pump water at a few houses in the 17th century.[14] However, the cost and complexity of steam engines meant that, even in the 19th century when steam engines were more efficient, their use for this purpose on country house estates was rare. At Cliveden, a steam engine pumped water from a well at White Place Farm on the west side of the Thames, via a pair of pipes running under the river, to the Italianate water tower (*see* Fig 3.9). At Tyntesfield the waterwheel-powered pumps that supplied river and well water to separate storage tanks on the hills above the house were supplemented after the First World War with an oil engine.

The arrival of electricity at the end of the 19th century greatly simplified the task of providing water to country houses and their estates. Drake and Gorham, who built a large steam-powered

Fig 3.9
The Italianate water tower, Cliveden. Water was piped to tanks in the tower from a farm on the far side of the River Thames by steam-powered pumps.

Fig 3.10
Part of the ruined
waterworks, Ascott, with a
water pump powered by an
oil engine.

electricity generator plant at Callaly Castle, Northumberland, in 1892, installed an electrically driven water pump at the same time in a former sawmill; much of the original equipment survives. Another remarkable survival is at Petworth House, where a well located in the basement, probably originally worked by a hand pump, had a three-cylinder pump driven by an underground donkey gin installed at some stage in the 19th century; this was then replaced with a DC electric motor in the early 20th century.

Until quite late in the 19th century, all these systems fed water from wells, springs or rivers, for use throughout the house and its estate, with no treatment other than some simple screening to remove debris. As levels of pollution rose, this clearly became a problem. Records show that the water supply at Audley End, which had hitherto been taken directly from the River Cam, was switched to a spring over 1km upstream around 1872.[15] More sophisticated purification systems were introduced in the 1880s, using porous ceramic pots, or sand and carbon filters which oxidised impurities.[16] Ascott, Buckinghamshire, had a sophisticated waterworks dating from the 1890s, with a purification plant to treat the river water and a Crossley oil engine which pumped it up to various tanks in the roofs of the house and outbuildings (Fig 3.10). At Wimpole Hall, Cambridgeshire, a building at the Wood Yard complex contains the remains of a water filtration plant with a triple-throw pump which was probably originally powered by a gas engine (the gasworks being next door), before this was replaced with an AC electric motor in the 20th century. The extensive improvements carried out at Polesden Lacey in 1903–4 included sinking a new well and borehole to a depth of 134m, from where a pair of electrically driven pumps supplied the house and estate.[17] The technology of water treatment was being pioneered by the rapidly expanding public water utilities in the second half of the 19th century and occasionally their networks extended far enough, even then, to supply country house estates; one example was Waddesdon, Buckinghamshire, which was supplied with water from its construction in around 1880 by the Chiltern Spring Water Company, which members of the Rothschild family had helped to establish.

Rainwater collection

Rainfall has, of course, been an important source of water for households for millennia. For more modest dwellings, this might have been nothing more than a water butt fed from a downpipe, while many larger houses had large underground cisterns to collect rainwater, as at Knole, in Kent.[18] As country houses became larger and more complex in the 18th and 19th centuries, so these arrangements became more sophisticated. Occasionally, rain from upper parts of roofs was collected in tanks within lower roof spaces, so that it could then be supplied by gravity within the house. More usually the collection tanks were below ground level, so the water had to be pumped back up to storage tanks if it was to be used within the house, although it could sometimes be supplied by gravity from underground tanks for use in the gardens. Rainwater tanks were mostly located under courtyards, often outside the kitchens or laundry, but at Ickworth the large tanks, complete with a filter, are sited in the basement of the house. Because of its softness, rainwater was prized for laundering: at Audley End, for example, a large rainwater tank with filters, dating from around the end of the 18th century, survives under the courtyard outside the laundry. The hand pump which lifted the water into a storage tank within the laundry has been reinstated as part of the restoration of the laundry block. Such features were still being included in new houses at least until the end of the 19th century: Wightwick Manor, a suburban industrialist's villa in the West Midlands, built around 1887, has a rainwater tank under the floor of its small laundry, with a hand pump to raise the water to a tank on the floor above. At Ickworth, the original rainwater hand pump was later supplemented by an electric pump.

In addition to using rainwater in laundries, some houses piped rainwater to other parts of their service areas, and sinks with a third tap, labelled 'soft', in addition to the usual hot and cold taps, survive. Examples include Dunham Massey (*see* Fig 3.29), where a large cranked hand pump lifted rainwater into a roof tank, and at Berrington Hall, Herefordshire. Around 1880, an automatic rainwater separator was invented, which diverted the first flush of water from the roof (which was likely to contain leaves and other debris) to waste before harvesting the remainder of the flow, but no surviving examples of these have been discovered.[19]

Water supplies within the house

Very few country houses had any provision for water supplies above the ground floor or basement before the second half of the 19th century, with some exceptions as noted earlier in this chapter. One use for piped water at these levels of the house was to provide a supply to 'buffets', elaborate marble basins usually sited in or near the dining room which could be used for washing glasses as well as for hand-washing. At Beningbrough Hall, Yorkshire, one of these 'buffets', dating from the early 18th century, was moved in the 1920s to a new position in Lady Chesterfield's bathroom after she had bought and renovated the house; the buffet had previously been supplied with water from the brick-built water tower on the banks of the River Ouse (Fig 3.11).

The Royal chambers at Westminster Palace were reportedly provided with piped hot water as early as 1352 and some monastic buildings were similarly blessed, but this was exceptional; it is not clear how this was achieved but it is assumed that metal tanks sited over open fires supplied the hot water by gravity.[20] The luxury of having both hot and cold running water throughout was not achieved in most country houses until the late 19th century and, in some cases, even later. Before then, hot water usually had to be carried in metal cans from a room at the bottom of the house such as the scullery, where it had been either heated in kettles on a range or, often, by a tank built into the range itself (*see* Chapter 6); the cans used for this purpose survive at many properties. The task of supplying hot water for washing and bathing in a large house with many guests was therefore

Fig 3.11
A marble basin, or 'buffet', moved to Lady Chesterfield's bathroom in the 1920s.

immense. Even as late as 1892 at Shugborough, Staffordshire, servants were carrying hot and cold water upstairs from a tap in the basement to every occupied bedroom in the house, a total of nine journeys per day for each room.[21] Audley End adopted a novel solution which made this task a little easier. A service area on the top floor of the house, known as the Coal Gallery (Fig 3.12), which was created around the 1780s, had piped cold water from tanks in the roof space above and stoves on which the water could be heated, so at least the servants were carrying the hot water downstairs to the bedrooms rather than upstairs; around the middle of the 19th century the stoves were replaced with a copper similar to those common in laundries.[22]

The provision of constant piped hot water was made possible by the development of the calorifier, a tank in which water was heated by a primary circuit, consisting of a coil of pipe, through which hot water was circulated by convection from a boiler located below. It is not clear precisely when these were introduced but it was probably around 1850, when the same principle was first adopted for space heating (*see* Chapter 5). Most early calorifiers were fairly crude in design, made on site by the plumber, and often provided hot water only to their immediate vicinity. Examples survive in the kitchens or sculleries at several properties, including Petworth House, where the primary circuit was heated by a nearby kitchen range; at Nostell Priory, Yorkshire, the calorifier in the kitchen was heated by a steam boiler. These systems were also often found in laundries, with the boiler primary circuit sometimes heating the drying cabinets as well as the calorifier.

Towards the end of the 19th century, centralised hot water systems became more common. In these, a single calorifier, heated by a boiler located in a distant boiler room, supplied hot water throughout the house. Such a system was installed at Audley End in the 1870s and around a decade later Tatton Park, Cheshire, installed a calorifier to provide hot water, with its primary circuit fed from a central steam plant (*see* Chapter 5). Calorifiers, albeit now with pumped primary circuits, continue to be used to provide hot water in most houses today. Instantaneous water heaters fuelled by gas were developed in the last quarter of the 19th century and quickly became popular in the houses of the lower middle classes, but no evidence has been found of their use in country houses until after the Second World War.

Fig 3.12
The Coal Gallery on the second floor at Audley End, with a 19th-century copper for hot water supplies and bins for storing coal, which was raised from the gardens below by a hoist.

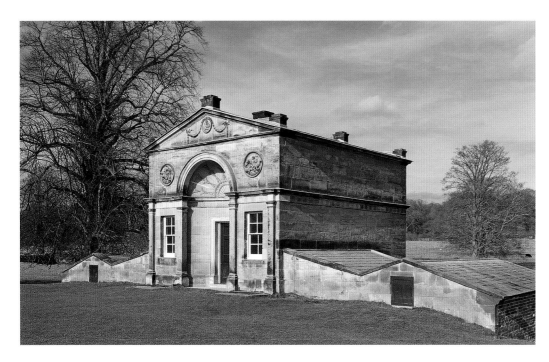

Fig 3.13
The Fishing Lodge at Kedleston Hall, designed by Robert Adam and built 1770–2, with a cold plunge bath in the basement.

Bathrooms and bathroom fittings

The concept of taking a bath to keep clean was not generally thought important before the 19th century, although people did take baths for health or recreational reasons at spas or in the sea by the late 18th century. At Penrhyn Castle, Gwynedd, a bathing pond complete with a pavilion was constructed in the early 19th century alongside the sea wall on the northern side of the park, although little trace remains now.[23] Cold plunge baths, again for health reasons, were built in the grounds or basements of some country houses for the benefit of gentlemen who wished to continue the bath treatment in their own homes or to enjoy some comfort after a hard day in the saddle. A very early example was built 1719–20 at Carshalton House, near Sutton, home of Sir John Fellowes, who was one of the directors of the South Sea Company. The marble-lined plunge bath was, unusually, contained within the brick-built water tower, inside which a waterwheel pumped water up to a cistern at the top.[24] Capability Brown designed the external plunge bath in the Gothic style at Corsham Court, Wiltshire, home of the Methuens, in the 1760s, while Robert Adam constructed a small, semicircular cold plunge bath fed by a spring in the basement of his Fishing Lodge at Kedleston Hall, Derbyshire, between 1770 and 1772 (Fig 3.13). At Wimpole Hall in the 1790s, Sir John Soane designed a conduit house, built

to resemble a Roman *Castello d'acqua* (an ornamental building looking like a small temple). It contained a tank that provided the water for a plunge bath on the ground floor of the house, near the chapel (Fig 3.14). Some sources have

Fig 3.14
The 18th-century plunge bath, Wimpole Hall.

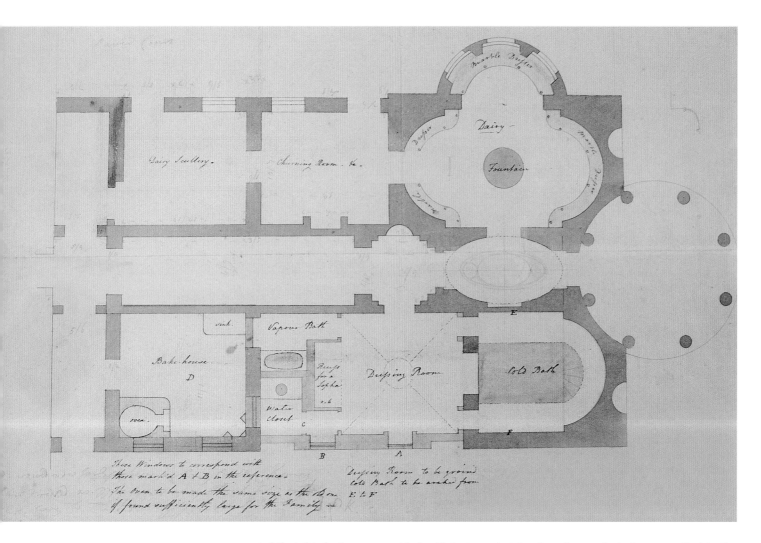

Fig 3.15
A plan of the bath house, Dodington, Gloucestershire; the building also housed the dairy and bakehouse.

suggested that this bath was provided with hot water but it has not been possible to verify this.[25] Plans for a classical-style building designed by James Wyatt for Dodington Park, Gloucestershire, around 1797 suggest it contained a large cold bath, a small 'vapour bath', which was presumably heated, and a drying area with a sofa, echoing the Roman approach to recreational bathing (Fig 3.15).[26] Curiously, the plans show this building as also housing the dairy and bakehouse; it was subsequently converted into the Dower House. When the 6th Duke of Devonshire employed Sir Jeffry Wyatville to make alterations to Chatsworth, Derbyshire, there was already a bath which Stephen Glover referred to as a swimming bath in 1830.[27] The Duke's *Handbook*, written in 1845, shows that the marble from this old bath was taken for use in the new bath, still referred to as a swimming bath in 1834.[28] He also referred to two smaller baths which 'are convenient and much frequented when I have company'. Both were heated by

what he describes as 'retorts over a fire' in the basement.[29]

Another fashion during the second half of the 19th century was for the provision of so-called Turkish baths in both public and private institutions and some country houses followed suit. These were not the traditional *hammams* of the Middle East, with their steam rooms, but generally consisted of one or more rooms of dry heat where the bathers sweated freely, followed by a cold plunge or shower and a cooling-off period in a cool room. Only two examples in country houses are known to survive, both in houses built originally for members of the nouveaux riches and now owned by the National Trust. Both have been restored for display purposes, so their original layout is not entirely certain. The earlier one, at Cragside, was designed by Lord Armstrong's architect, Norman Shaw, as part of his first additions to the original house in 1870. Shaw may well have been assisted by Tyneside architect James Shotton, who had already built several Turkish baths.[30] The bath suite, now consisting of a hot room, a cooling room and a plunge pool and separate shower, appears from his drawing to have undergone considerable modification during its construction (Fig 3.16). It was cleverly situated between chambers with huge water pipe coils which were the source of the hot air that was ducted up into the main house (*see* Chapter 5). The suite was therefore warmed through the walls on both sides.

The other example of a Turkish bath was constructed at Wightwick Manor, near Wolverhampton, for Theodore Mander in 1887. The plans, again, suggest modification during construction, the original layout consisting of two hot rooms and a cooling room with no provision for a bath, as there is now; it is not known when this was added. The system was heated by what was described in *The Architect* shortly after the house was built as one of Messrs J Constantine's 'Convoluted Stoves', a type of heat exchanger to produce the necessary high temperatures for the hot room; the remains of this are still in situ.[31] At Mount Stuart, on the Isle of Bute, the plans drawn up for the Marquess of Bute's rebuilt house show that in 1879 a Turkish bath and shower room were certainly planned adjacent to a large heated swimming pool in the basement. These were described in a newspaper report of 1885:

> The swimming bath measures about 30 feet by 25 feet and has a groined roof of terracotta and supported by Aberdeen granite columns. The floor of the bath slopes so that at one end it is about 2 feet 6 inches deep and at the other end it is about 7 feet or 8 feet deep. A smaller dressing room in connection with the bath is lined with tiles, and groined in the ceiling with terra cotta. The Turkish Baths which adjoin have groined concrete ceilings with stone ribs, supported at regular intervals by stone pillars.[32]

How far the Turkish baths were ever completed to serve their purpose is uncertain but the rooms are still there; the swimming bath was brought back into use by the present Bute family in the 1970s and still makes use of the original water circulation system.

Baths such as these were, however, the exception to the rule and were used more for recreation

Fig 3.16
The plunge bath in the Turkish bath suite at Cragside.

DOMESTIC SANITARY REGULATIONS.

Fig 3.17
'Domestic Sanitary Regulations', by John Leech, Punch, January 1850.

than cleanliness. In most cases, people washed in their bedrooms or dressing rooms, using portable equipment such as basins, ewers and hip baths, which were filled and emptied by the servants. Early showers, too, were essentially portable in that they were free-standing, and were filled and emptied by servants. These were fairly common by the middle of the 19th century, when a cartoon by the famous political cartoonist John Leech appeared in *Punch* in 1850 showing reluctant children being led towards such a shower (Fig 3.17). As can be seen, the shower had a very limited capacity but several examples do still survive in country houses, including Erddig, Wrexham (Fig 3.18), Wimpole Hall and Calke Abbey, Derbyshire. Generally, though, the bathroom as a separate room remained a rarity in country houses, compared with middle class homes, before the final decades of the 19th century, despite Robert Kerr's assertion in 1864 that 'No house of any pretensions will be devoid of a general bathroom.'[33] The availability of servants in country

houses was partly responsible for this, as suggested in Chapter 1. Recalling her years at Longleat, Wiltshire, in the 1930s, and what she had been told about this great house prior to 1914, Daphne Thynne, wife of the 6th Marquess of Bath, remembered nostalgically that:

A housemaid's work must have been made very arduous by the constant carrying of hot-water jugs and the laying out and clearing away of the hip baths, which were placed in the bedrooms in front of the fire. There must have been a far greater comfort in the cosy sight of a copper hot-water jug nestling in a warm, white towel and reflecting in the bright lights of the blazing fire than there is in the ugly modern luxury of hot and cold water laid on in the bedrooms.[34]

Similar nostalgia for the pleasure of bathing in one's own warm room rather than going down the corridor to an often chilly bathroom can be found in many accounts of life in country houses

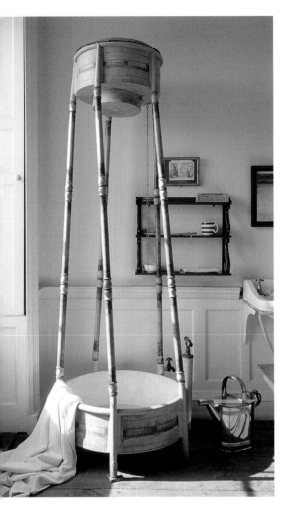

Fig 3.18
*A free-standing shower,
Erddig.*

from the late 19th and early 20th centuries, and this probably deferred the introduction of bathrooms for some time after they were technologically possible. The dressing room adjacent to the bedroom was often chosen as the location not just for the traditional hip bath, but also for the new baths where water was heated by a charcoal or coal fire or by gas. A wonderful example of one of these can be seen in the Silk Dressing Room at Tatton Park, a tin bath on castors which could presumably be wheeled near to the fireplace to provide the outlet for the flue (Fig 3.19). Maids had then to both fill and empty the bath and stoke the firebox to warm the water! Gas-heated baths were introduced from the 1860s, one of the leading makers of these being John Wright and Co of Birmingham. Unlike the coal-heated baths, however, these could not be wheeled around as their position was fixed by the gas supply pipe and they usually had fixed cold water supplies and a waste pipe. Since few country houses tended to have gas laid on to the upper floors (*see* Chapter 4), they cannot have been common and seem to have been confined to middle class houses.[35]

The addition of separate bathrooms to country houses became more common from the 1870s, although this was dependent on the ability to pump hot water to the upper floors. At Powis Castle, Powys, the 4th Earl employed George Frederick Bodley to modernise the castle

Fig 3.19
*The tin bath in the Silk
Dressing Room at Tatton
Park. The bath was filled by
maids with hot water and it
was kept warm by a firebox
at the bottom.*

following his accession to the Earldom in 1891, and he managed to insert bathrooms into odd spaces on the upper floors of the house. Far more drastic modification was undertaken by George, 7th Lord Rodney, at Berrington Hall, after his marriage to Corisande Guest in 1891 (*see* Chapter 1). Lord Rodney had a three-storey rendered brick tower built onto the courtyard facade of the house which had a water tank in the roof space and bathrooms on each floor. A rather ugly structure, this was demolished by the National Trust when they acquired the house in the 1950s.

Most of the early full-length baths were made of cast iron lined with vitreous enamel and

Fig 3.20
A wood-panelled bath equipped with an overhead shower, Tyntesfield.

enclosed in wooden panelling to hide the fittings and pipework. Several of these survive in country houses, such as that in the late 19th-century Nursery Bathroom at Lanhydrock, Cornwall. The wood-panelled bath in a bathroom on the first-floor guest corridor at Tyntesfield has an early plumbed-in shower above it, probably dating from the 1880s (Fig 3.20). At the other end of this bath from the bath taps is a mixer tap for the shower supply, with a test cock for checking the temperature. Once the shower tank was full, a chain was pulled to release the water. Such fitments were soon succeeded by canopy baths, with an enclosure fitted with a shower at one end of a full-length bath. Most shower enclosures were wood-panelled, but at Tredegar House, Newport, there is a glass-and-metal shower enclosure at one end of the wood-panelled bath in the so-called Master's Bathroom, remodelled *c* 1905 for the newly created Viscount Tredegar.

There was considerable criticism of panelled baths, as they were difficult to clean; German architect Hermann Muthesius stated categorically (but perhaps not wholly correctly) in 1904, 'In good bathrooms in the past the bath used to be regularly encased in wood; but the custom has now ceased entirely and all parts are expected to be accessible for cleaning.'[36] Consequently, free-standing cast-iron baths, often ornamented with rolled rims and claw feet, became popular. Many of these were made by sanitary-ware manufacturers such as Shanks, Twyford or Doultons, who added iron foundries to their premises. Examples of their products, dating from the early 20th century, can be seen in many country houses as bathrooms became more of a standard feature. The combined bath and shower in Julius Drewe's bathroom at Castle Drogo, Devon, was one of seven chosen by him from the sanitary-ware manufacturer John Bolding in the second decade of the 20th century (Fig 3.21). By this time, washbasins were also being installed into bathrooms, and sometimes into bedrooms or dressing rooms, often set into wooden or marble cabinets or standing on metal legs. One, also made by Bolding, with an elaborate tiled splashback, is in the same bathroom at Castle Drogo as the shower bath, while an earlier one of marble made by Royal Doulton in about 1905 can be seen at Tredegar House (Fig 3.22). The sanitary-ware manufacturers began to introduce all-ceramic pedestal washbasins by about 1910.[37]

Muthesius praised England in 1904 for having led all the Continental countries in developing

Fig 3.21
A combined bath and
shower made by John
Bolding, at Castle Drogo.

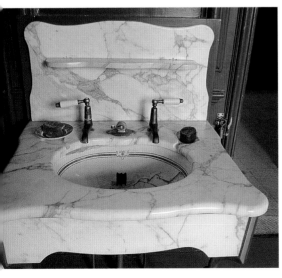

the bathroom, although he was probably refer-ring more to middle class homes than country houses.[38] However, he also claimed, 'It is alien to the nature of an Englishman of standing to envelope [sic] himself in luxury,' and the bath-rooms surviving in country houses from the early 20th century would seem to support his point of view.[39] The elaborate decoration of the Victorian period was succeeded by a desire for simplicity, probably linked to an increasing consciousness of the link between cleanliness and health. Most baths and washbasins were being made of white earthenware or, in the case of baths, cast iron lined with white enamel. From the late 1920s, it became fashionable to have coloured fittings, although this was an American import to which not all subscribed. Tiles had

Fig 3.22
A marble washbasin made
by Royal Doulton at
Tredegar House.

Fig 3.23
Virginia Courtauld's
magnificent bathroom at
Eltham Palace, designed by
Peter Malacrida, c 1935.

ble washbasin and a marble bath, complete with gold-plated taps and a lion's mask spout, set within a gold mosaic niche containing a statue of the goddess Psyche (Fig 3.23). The provision of en-suite bathrooms for all the family and guest rooms at Eltham Palace is indicative of just how far the concept of the bathroom in country houses had come in half a century, but it was only made possible by the technological development of a hot water supply able to satisfy the needs of the family and their household.

Sanitation

Given the problems presented by the disposal of human waste from the large households of country houses, it is perhaps surprising that greater attention was not paid to the development of adequate sanitary facilities rather earlier than proved to be the case. The first lavatory with working parts and flushing water was invented by Sir John Harington (1561–1612), godson of Elizabeth I, in 1596. This does not seem to have been widely adopted and external earth closets, often in a remote corner of the garden, remained by far the most common toilets, together with the use of chamber pots and close stools inside the house. However, some country house owners did think it worthwhile installing these early types of flushing WCs. Thomas Coke, or his wife Margaret, had sufficient confidence to plan water closets throughout Holkham Hall in the 1730s, even in positions where there was no external ventilation. They were installed by Devall, a specialist from London. Nine were installed altogether, nearly as impressive as the ten water closets installed much earlier at Chatsworth.[41] Alexander Cummings took out a patent for an improved water closet in 1775; his design incorporated a sliding valve in the base of the pan operated by a pull-up lever housed in the toilet seat. The slider valve allowed the waste in the pan to pass through but then prevented its return or sewer gas leaking upwards. His invention was improved upon by Joseph Bramah, a cabinetmaker by trade who fitted up water closets. By 1778, Bramah had patented his own version of a valve closet, replacing the slider with a drop-down valve which was cleansed every time the pan was flushed (Fig 3.24). This proved so successful that he claimed to have made 6,000 closets by 1797, and they remained the first-choice WC for the well-to-do until improved designs were introduced in the second half of the 19th century.[42]

embellished bathrooms since the mid-19th century, but in the late 1920s new nonporous wall-coverings made an appearance. The late-1920s renovations at Upton House, Warwickshire, designed by the architect Percy Morley Horder, included an Art Deco-style bathroom for Lady Bearsted, with aluminium leaf surfaces.

Vitrolite was an opaque pigmented glass first developed by the Pilkington Company of St Helens, which could be cut into large tiles.[40] This was used in the bathroom designed for Stephen Courtauld at Eltham Palace, Greenwich, in 1935–6. His wife Virginia's bathroom was even more elaborate. It was designed by Peter Malacrida and had walls lined with onyx, a mar-

Some country house owners, not surprisingly, took rapid advantage of this development. One of the earliest must have been a new recruit to the gentry, Sir John Griffin Griffin, who inherited Audley End from his aunt, the Countess of Portsmouth, in 1762. In July 1775 a 'patent apparatus for a Water Closet' was ordered from Joseph Bramah at a cost of £8 8s, together with 'pipes to cistern, stink trap, service box, cover, brass hammer soldered, all complete'. This order was therefore placed about three years before Bramah patented his design. Another guinea was paid out for a man to go down to Audley End for four days to fix the apparatus.[43] The total cost was £16 11s 6d, which was to include a personal visit from Bramah himself, although the document implies that he did not attend personally, but left it to his fitter. The experiment seems to have been successful, for by October 1785 another four patent WCs were ordered from Bramah and sent to Audley End from the Dolphin Inn, Bishopsgate, by carrier.[44] The invoice was written and signed by Bramah himself, for a total cost of £36 17s, a not inconsiderable expense at the time, equivalent to nearly four thousand pounds today. Audley End, therefore, appears to have had five Bramah WCs installed by 1785, only seven years after Bramah had

taken out his patent. It is more difficult to establish where they were placed in the house, but floor plans dating to 1787 suggest that locations on various floors of the house were chosen, including a space beneath the main staircase from the main hall, dressing rooms on the first floor and in a small alcove near the back entrance to the house on the north side, the latter presumably for the use of the servants who had no access to close stools within the house.[45] Another early purchaser was George Adams Anson of Shugborough Hall, nephew of Admiral Anson, who purchased two Bramahs in 1780 at a cost of about £103.[46] His servants, however, had to continue to use earth closets situated in the stable block. The Earl of Moira, friend of the Prince Regent, had at least six water closets installed on two floors in his home at Donington Park in Leicestershire by 1813.[47]

Several sanitary engineers patented improvements to Bramah's design of a valve closet by the middle of the 19th century, including J G Jennings, who supplied WCs for the Great Exhibition of 1851.[48] These still needed the support of a wooden frame, which also disguised all the various fittings, but the pan was by now ceramic rather than metal. Such valve closets continued to be made throughout the 19th

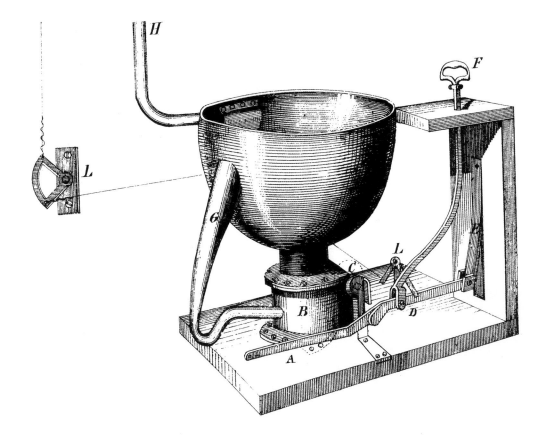

Fig 3.24
Bramah's flushing WC;
the mechanism was housed
in a wooden case which
concealed the pipework.

century, and one with a ceramic bowl embellished with a design of blue flowers was installed by Jennings for the Marquess of Bute at Cardiff Castle in the 1870s. The valve closet was flushed, often rather ineffectively, by water from a spout placed under the lid. In 1870, Dent and Hellyer, sanitary engineers of London, patented the first pans in which the water was circulated all round from below the rim, which was much more effective. Examples of their 'Optimus' model can be found at Audley End (Fig 3.25), where the descendants of Sir John Griffin Griffin continued to improve their sanitation, and at Brodsworth Hall, Yorkshire. When George Devey extended Ascott House for Leopold Rothschild in the 1870s, he installed several of John Bolding's valve closets in mahogany, together with a sewage treatment plant made by Thomas Crapper.

Perhaps the most revolutionary change in the design of WCs in the second half of the 19th century was the adoption of free-standing ceramic pedestals which did not need the support of wooden frames as previously. These were produced for various firms of sanitary engineers by the large potteries of Staffordshire and South Derbyshire, Twyford and Shanks probably being the best known, together with Doultons of Lambeth in London. The two major types which developed were the wash-out and the wash-down. The wash-out WC needed a considerable flush from the back to the front of the bowl, which then passed into the trap, so wash-out WCs have a bulge at the front of the pedestal. An elaborately patterned wash-out WC by Doultons can still be seen at Saltram House, Devon (Fig 3.26). The wash-down had the trap towards the back and also retained water in the bowl, the pedestal being straight at the front; it was similar to modern WCs and became the choice of

many from the 1880s onwards. Among the many models of ceramic WCs produced, two of the most popular were the 'Deluge', made by Twyford, and the 'Invictas', made by Johnson Bro's of Hanley (Fig 3.27); examples of the former can be seen at Knightshayes, in Devon, and Wightwick Manor, together with the set of underground servants' toilets placed off one of the service tunnels at Alnwick Castle, Northumberland. Many late-Victorian models were heavily decorated with transfer printing or made in embossed earthenware, but in the early 20th century much plainer and often white sanitary ware became the most common type used.

As WCs became more common, the water companies became concerned about the amount of water used in flushing them, and so various Acts of Parliament were passed in an attempt to prevent waste of water. Numerous patents were taken out for cisterns designed to allow only a limited flush of water, including the use of the ballcock. Even so, it was possible for users to terminate the flush prematurely and so fail to clean the toilet pan. A solution to this was the siphonic cistern, where a set amount of water was released using siphonic action, regardless of the length of time the chain or handle was pulled down. The development of the low-level cistern helped to minimise the noise of this flush and these were developed from the 1890s, using the wash-down pan rather than the wash-out. They became known as 'combination closets' and were first developed by the firm Shanks, although Doultons and others soon came to make them too.[49]

THOMAS CRAPPER & COMPANY,

Pedestal Wash=down Closets.

The "Invictas."

Combination No. 5.

The "Lennox."

Combination No. 6.

No. 5 Combination, comprising White and Printed "Invictas" Closet, with Fig. 211 Polished Mahogany Seat and Back Board, and 2 Gallon Fig. 219 Water Waste Preventer with 1½ in. Fittings, and China Pull and Brackets ... £3 6 3

No. 6 Combination, comprising Strong Cane & White Fire-Clay "Lennox" W.C. £1 0 6
Fig. 211 Polished Mahogany Seat with Back Board 1 1 3
2 Gallon Fig. 814 Valveless Water Waste Preventer with Cover, and China Pull 1 2 3
Iron Brackets 0 1 0

£3 5 0

Fig 3.28
A urinal adjacent to the
billiard room, Tyntesfield.

the cloakroom behind the billiard room at Tyntesfield contains one of the few surviving urinals found in country houses, made by Jennings (Fig 3.28). The cloakrooms adjacent to the entrance hall at Eltham Palace were supplied with urinals made by Armitage Shanks, and Boldings's 'Cavendish Syphonic Pedestal Action' WCs, that firm also having supplied most of the sanitary fittings in the house, some of which still exist.[51]

Sewage disposal

The contents of chamber pots were traditionally collected for use in the gardens or for other purposes, and Calke Abbey has a chute believed to have been installed for this purpose, leading from the first floor to a container outside the back door. Many households made provision for the disposal of slops from bedroom chamber pots and basins in specially constructed sluices in small rooms off the back stairs. They were usually well hidden from public view, but examples can be seen at Dunham Massey (Fig 3.29), Lanhydrock and Wightwick Manor. Drainage culverts carrying waste water out of the house existed in the medieval period and houses reconstructed from monastic remains could take advantage of these, while other houses had some drainage provision included in their original design. Once flushing WCs were introduced, the design and construction of drains had to be improved, with traps to avoid foul odours seeping back into the house. The Bramah WCs installed at Audley End in the 1780s appear to have discharged into a sewer system that survived from the Jacobean mansion, and also had an unusual facility which allowed water from a pond in the grounds to be diverted into the sewers periodically to flush them clean and also allowed them to be used as a fire hydrant system.[52] This sewer discharged directly into the River Cam until 1904, when a pump driven by an oil engine was installed to pump the sewage from a holding tank to a disposal ground on a nearby hillside, where it was allowed to percolate into the ground through a bed of porous material. The sewer from WCs installed at the time of the original construction of Holkham Hall in the 1730s discharged directly into the nearby lake.[53] Such arrangements were common – if there was any sewage treatment at all it was limited to collecting the sewage in cess pits, where limited bacterial action took place to break down the waste. Solid material could also be separated out and treated with lime before being used as fertiliser. The rela-

Despite all these developments in the design of WCs in the second half of the 19th century, their adoption in country houses was comparatively slow, many ladies still preferring to use the more private facilities of close stools in their own rooms. WCs were commonly installed in the gentlemen's wings that were added to many houses, which included billiard rooms, gun rooms and cloakrooms: Kerr recommended the installation of a WC in the cloakroom and suggested, perhaps ironically, 'The reason for having these conveniences connected with the Entrance is that they are provided for the use chiefly of gentleman visitors, who can always find their way to the Entrance-Hall without trouble, if nowhere else.'[50] A good example of a gentleman's cloakroom survives at Wightwick Manor, while

Fig 3.29
The 1906 slop sluice,
Dunham Massey, with
original taps, including one
for rainwater.

tively low-lying location of Blickling Hall, Norfolk, meant that sewage and surface water regularly flooded the basement, a problem which was overcome in the 1890s with the installation of a sewage pump, powered by a waterwheel, which still exists under a manhole cover in the gardens.

More scientific approaches to sewage treatment did not emerge until the very end of the 19th century, with the development of the septic tank, in which anaerobic bacteria broke down and sterilised the waste more completely than was achieved in a cess pit, followed by the contact bed, which treated liquid waste by passing it through a filter of coarse material where oxidation killed the remaining harmful bacteria.[54] The engineer A Alban H Scott wrote in 1911: 'The old system of discharging the whole of the sewage into a cesspool or midden, and allowing the liquid to gradually filter into the adjoining land, though a convenient form of "disposal", is hardly such as will be accepted today.'[55] By this time, the typical municipal sewage plant consisted of a coarse screen to remove large solid material, followed by septic tanks and contact beds filled with coke, furnace slag or other rough-surface material. Relatively few country houses invested in such sophisticated installations but those which did included Cliveden and Polesden Lacey.[56] At Ascott, Thomas Crapper & Co of London built a sewage works around 1899, which is now disused (Fig 3.30). The remains

of another example, also dating from the late 1890s, were excavated recently at Canwick Hall, Lincolnshire (Fig 3.31).[57] Later in the 20th century, some country houses were able to connect to municipal sewage networks but a great many retain their own sewage plants to this day, either conventional systems or, as at Dudmaston Hall, in Shropshire, reed bed filtration systems.

Fig 3.30
The derelict late 19th-century sewage works, Ascott, supplied by Thomas Crapper and Co. The building housed the coarse screen and staff facilities; in front of this are the septic tanks and the contact bed filters are in the foreground.

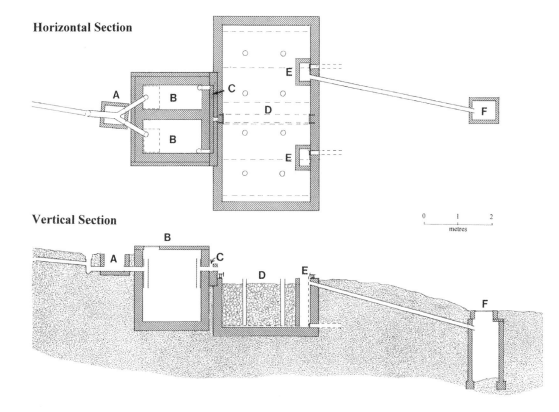

Horizontal Section

Vertical Section

0   1   2
metres

Fig 3.31
A plan and section of the sewage works, Canwick Hall, Lincolnshire.

Laundries

From medieval times, country houses had had dedicated rooms for the washing of linen but, as with the spaces devoted to most other household tasks, these became increasingly sophisticated from the 18th century onwards. The laundries handled all the bed and table linen and most of the clothes of the household, including the staff, although the more delicate items of clothing were sometimes washed by the owners' personal servants. Country house laundries often also took the bed and table linen from the families' London houses from the second half of the 19th century, when the arrival of the railways made transportation easier. This was the case at Chirk Castle, Wrexham, where the so-called 'bothy laundry', situated some way from the house, received regular consignments of dirty laundry all the way from London.[58] The classic layout of a laundry, described by authors such as

Kerr, consisted of a wet laundry or wash house for washing, an adjacent drying ground and a dry laundry for indoor drying and ironing; ideally these rooms would have high ceilings and good ventilation.[59] In some cases, they were built one above the other in double-height rooms, as at Castle Ward, County Down, which has a gallery supported on wooden columns on all four sides. Initially, these facilities were located in service areas adjacent to the house but in the latter half of the Victorian period, concerns grew for the moral welfare of the laundry maids. At Hesleyside in the 19th century there were 'scandals in the laundry, which ... was nothing but a brothel until a new entrance was built ... The upper servants were among the most licentious.'[60] As a consequence, new laundry complexes, with their own self-contained accommodation for the staff, were often built in distant parts of the estate; examples of these include Dudmaston Hall, Stokesay Court (Fig 3.32), Culzean Castle, in Ayrshire, and Lyme Park, in Cheshire.

Wet laundries generally contained at least one copper for boiling clothes and, later in the 19th century, a separate boiler for providing hot water for the sinks (Fig 3.33). The sinks themselves were usually made of wood or stone, often with sloping fronts or sides to facilitate the use of washboards, and were located under the windows for maximum light (Fig 3.34). As well as conventional roller mangles, wooden box mangles were used to extract most of the water from larger items such as sheets and tablecloths before they were dried outdoors, often in secluded drying grounds, or on racks suspended from the ceiling of the dry laundry (Fig 3.35). A later refinement was the introduction of heated drying cabinets, with racks which slid out on

Fig 3.34
Sinks with sloping fronts, in the wet laundry, Killerton.

Fig 3.35
The double-height dry laundry, Beningbrough Hall, with drying racks, a box mangle, ironing table and a stove for heating flat irons.

Fig 3.36
*Drying cabinets in the dry
laundry, Berrington Hall.*

Fig 3.36
*Drying cabinets in the dry
laundry, Berrington Hall.*

Fig 3.37
*A solid fuel stove for
heating irons in the dry
laundry at Castle Ward.*

rails (Fig 3.36). An early version of one of these, heated by a small stove alongside, still exists at Kingston Lacy, Dorset; later, commercially-made versions can be seen in many country house laundries, including those at Erddig, Berrington Hall, Stokesay Court and Killerton, in Devon.[61] The other key components of the dry laundry were the large wooden tables on which laundry was ironed and the stoves on which flat irons were heated. Sometimes these stoves were little different from normal solid fuel ranges but special stoves were also developed which enabled a large number of irons to be heated rapidly; the most common of these was the 'Pagoda' design, examples of which survive at Dunham Massey and Florence Court, while there is a massive version by an unknown maker at Castle Ward (Fig 3.37). Ironing with flat irons was an arduous task, so it is no surprise that one of the first electrical appliances to come into widespread use was the electric iron.

Mechanisation to assist the hard physical work in the laundry was slow to be adopted, perhaps because early washing machines were so ineffective and often violently resisted by laundresses proud of their skills.[62] From around 1860 companies such as Thomas Bradford & Co of Salford manufactured machines which agitated

Fig 3.38
The mechanised laundry,
Holkham Hall; line
shafting powered by an
electric motor drove a
washing machine, spin
dryer and mangle.

the clothes in a horizontal wooden drum, but they had to be filled with water and emptied by hand, and were usually cranked by hand, too. A small number of large, forward-thinking estates, including Holkham (Fig 3.38), Blenheim and Tatton Park, introduced mechanisation, with steam engines powering washing machines, mangles and spin dryers (known as 'hydro-extractors') via line shafting. Tatton's laundry, powered from the steam engine which also drove a refrigerator and electricity generator, included an 'ironing machine' but it is not clear how this operated.[63] Even after the arrival of electricity at the end of the 19th century, mechanical aids were slow to arrive in the laundry, mainly because the equipment then available was so much less efficient than the traditional methods. Towards the end of the 19th century, the growth of large mechanised commercial laundries in some cities and larger towns provided an alternative means of washing and ironing the bulk linen for a few estates.

In conclusion

It is evident, then, that water supply was the first area of domestic life on which technology made a significant impact, with water- and animal-powered pumps, first adopted in mining industries, supplying water for both garden features and domestic use on many estates from the early 18th century and sometimes earlier. The provision of piped cold water supplies within the house must have eased the workload of servants significantly but this process was only complete when piped hot water supplies became common, which was often not until quite late in the 19th century. Piped cold water supplies also paved the way for the introduction of the WC, which was genuinely a convenience for all, although the use of portable sanitary equipment continued as long as there were personal servants available to service private bedrooms. A splendid variety of different sanitary fittings survive in country houses, which are a source of great fascination for most visitors today, as are the many laundries which have been restored for public display. As with many restorations, it is not always easy to distinguish original from replacement equipment, but even with this in mind, these laundries – perhaps more than many other aspects of domestic activity before about 1900 – demonstrate the huge burden of servants' work before the introduction of modern equipment.

Lighting and energy production

'Suddenly, at the turning of a screw, the room was filled with a splendour worthy of the palace of Aladdin.'

J G Lockhart, 1853[1]

Lighting was an area of new technology in which country house owners often made significant investment, particularly in the 19th century. This is unsurprising, because developments in artificial lighting from the end of the 18th century – firstly in improved oil lamps, then gas lighting and finally electricity – brought significant and immediate improvements to home life, as they did in industry and commerce, at least for those wealthy enough to afford them. However, even with these improvements, artificial lighting remained an expensive commodity, at least until the 20th century. As we have seen with water supplies and sewage disposal, the remote location of most country houses placed them beyond the reach of the early utility networks, forcing them to become self-sufficient in gas manufacture and, later, electricity generation, sometimes into the mid-20th century.

Artificial lighting before the industrial age

From the perspective of our modern, 24-hour society, it is hard to appreciate the changes that developments in artificial lighting brought. For most of man's existence, his activities have been largely dictated by the sun.[2] Joseph Swan was, as described below, one of the pioneers of electric lighting and even though he was born in 1828, when gas lighting was becoming widespread in Britain, he neatly described the experience of the majority of the population:

The days of my youth extend backwards to the dark ages, for I was born when the rushlight, the tallow dip or the solitary blaze of the

hearth were the common means of indoor lighting … In the chambers of the great, the wax candle, or exceptionally, a multiplicity of them, relieved the gloom on state occasions; but as a rule, the common people, wanting the inducement of indoor brightness such as we enjoy, went to bed at sunset.[3]

As this quotation shows, the open fireplace remained an important source of light for many households into the industrial era. Burning wood torches were also commonly used in country houses long after the end of the medieval period, particularly for external lighting and in large internal spaces such as entrance halls. However, any surviving fittings which appear to have been designed to hold wooden torches are likely to be later reproductions and this style of fitting was also copied for some early electric light fittings.[4] Rushlights, made by dipping

Fig 4.1
A wall sconce with candles, Uppark, West Sussex.

rushes in animal fat, were common in humbler households but the predominant form of country house lighting until the late 18th century was the candle. In most cases, these too used animal fat – tallow – as the fuel. From the 15th century, the Tallow Chandlers Company controlled the manufacture of and trade in tallow candles – initially just in London but later their control spread more widely. The candles were also subject to taxation from 1709 until 1831, which rendered them unaffordable for most people. Beeswax candles, also taxed and subject to monopoly supply, were even more expensive; even in the wealthiest households, these were reserved for the grandest rooms and special occasions. Only in the second half of the 19th century, after the ending of taxation and the introduction of paraffin from America, did candles come in reach of the general population. As well as being expensive to buy, candles required the attention of staff to trim the wicks, so a house brilliantly lit with candles on a grand occasion was a very conspicuous demonstration of wealth.

Some country houses evidently made their own tallow candles from animal fat recovered from the kitchen. Castle Coole, County Fermanagh, has a Tallow House where candles were made, near the entrance to the service tunnel. The household accounts for Audley End House, Essex, from the 1830s show entries for the purchase of wax candles but not tallow ones; credits in the kitchen accounts for tallow suggest that they, too, made their own candles.[5] The high cost of candles meant that their use by servants was strictly controlled. Lower servants, at least, were generally not allowed them in their bedrooms, partly for reasons of fire safety and partly to discourage them from reading.[6] To light their way to bed they therefore had to steal the stub ends of candles, for which there was much competition, as butlers and footmen often regarded these as a 'perk', to be sold back to local merchants.[7]

Candle fittings, usually in the form of ceiling-mounted chandeliers, free-standing candelabra and wall-mounted girandoles and sconces (Fig 4.1), can be found in almost every country house, although many have subsequently been converted to electricity. Mirrors were often placed behind candelabra and sconces to reflect more light into the room; many examples were listed in the inventories at Knole, Kent, for example.[8] Devices such as water-filled bowls, sometimes known as 'lacemaker's condensers', were also employed to intensify the light. Despite these measures, candles offered an expensive and inadequate means of providing light for large spaces or for intricate work. However, candlelight was widely regarded as more congenial and flattering to ladies' complexions than the gas or electric light which followed, so the tables for grand dinners have often continued to be lit by candelabra to this day (Fig 4.2).

Simple oil lamps, which date back to prehistoric times, operated on the same principle as the candle, with a wick feeding oil to an open flame by capillary action, and produced a similarly poor quality light. Historically, these were more popular in regions such as the Mediterranean, with abundant supplies of vegetable oil such as olive oil. They became more common in Britain with the arrival of whale oil in the 18th century.

Fig 4.2
A dining table with candelabra, Attingham Park, Shropshire.

Argand oil lamps

The Argand oil lamp has been described as the 'first significant advance in lighting technology for millennia'.[9] It was invented in France around 1780 by the Swiss Ami Argand, and had a hollow cylindrical wick enclosed in a glass chimney supplied by gravity from a raised oil reservoir (Fig 4.3). The improved air supply to the wick increased the temperature and thus the brightness of the flame, producing a light equivalent to about ten candles.[10] Argand's invention enjoyed limited success in France but he came to England in 1783 and set up an initial partnership with Boulton and Watt, who manufactured the metal bases, reservoirs and burners of the lamps, and the London glassmaker William Parker, who made the glass shades and chimneys and marketed the finished product.[11] These lamps, either free-standing or suspended from ceilings, and often with multiple burners fed from a single reservoir, quickly became popular in wealthier households and Argand soon had to take on other partners to keep up with demand.[12] However, the numbers produced by official partners were dwarfed by the tens of thousands of utilitarian pirated versions made for use in factories.

Household accounts for Audley End show that one free-standing lamp was purchased from William Parker in 1785, at a cost of 18s; this evidently found favour as a further 22 were bought over the next year.[13] A pair of these survives in the first floor dining room, complete with their glass chimneys, which is rare since these were easily broken (Fig 4.4). Many country houses retain examples of multi-burner ceiling- or pedestal-mounted Argand oil lamps, although these have mostly been converted into electric lights. An example of one of many such lights which survive at Penrhyn Castle, Gwynedd, can be seen in Figure 5.15.

While Argand-type oil lamps offered significant advantages over candles and simpler oil lamps in terms of the quality of light they provided, they were still expensive to run and absorbed a significant amount of staff effort. Most types of whale oil gave off a smell of fish when burned, but this was less pronounced with the most expensive sperm whale oil. When demand for this started to outstrip supply, an alternative was found in the form of the equally expensive Colza oil, which was extracted from rape seed. This was supplanted later in the 19th century by various forms of mineral oil. Where possible, lamps were taken to a dedicated lamp room for filling, cleaning and for the wicks to be trimmed. Several examples of lamp rooms survive in country houses, including Florence Court, in County Fermanagh, Calke Abbey, in Derbyshire, and Penrhyn Castle; they can often

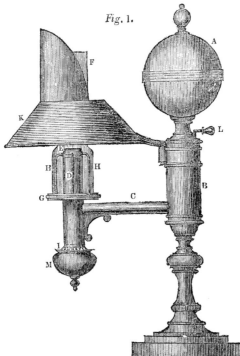

Fig 4.3
An Argand oil lamp. From The Penny Cyclopaedia, *Vol II, 1834.*

Fig 4.4 (far right)
A 1780s Argand oil lamp, Audley End, probably supplied by Argand's licence-holder, William Parker of London.

be identified by stains on the floor and a pervasive smell of oil. The lamp room at Castle Coole, which still retains an earlier plunge bath, boarded over, has a metal-topped counter with a drain hole, so any precious oil spilled during the filling of lamps could be recovered (Fig 4.5). Later in the 19th century, simple, inexpensive paraffin lamps designed to fit into candle holders, known as 'peg lamps', were developed. Examples survive at The Argory, County Armagh, and Erddig, Wrexham (Fig 4.6), the latter mounted on reclaimed gas lamp brackets. Erddig enjoyed little modernisation from the early 19th century until the mid-20th century and was briefly lit with improvised propane gas lamps before electricity was brought in after the National Trust took over the property in 1973.[14]

Lighting with coal gas

Throughout most of the 18th century, scientists in many parts of Europe experimented with the distillation of coal, wood or other organic material to produce a flammable gas. Most people regarded this as little more than a novelty, but William Murdoch, an engineer working for the Boulton and Watt Company, was the first to demonstrate the practical application of this idea, first in his house in Redruth and then in lighting parts of Boulton and Watt's Soho Foundry in Birmingham.[15] This gained the attention of textile manufacturers, major customers of Boulton and Watt, who saw it as a solution to the problem of lighting their factories safely and economically, and the world's first practical gas lighting was installed in two cotton mills in the north of England around the end of 1805. The method of making gas from coal was essentially simple: coal was heated in a closed vessel, called a retort, and the resultant gasses were cooled in a condenser to remove tar, and then passed through a purifier to remove impurities, before being stored in a water-sealed gas holder (Fig 4.7). The small gas holders employed at country house gasworks generally had below-ground brick-lined water tanks.

The basic technology was not patented, so the idea spread rapidly, not just in the form of small private installations for individual users like factories, but also from public gasworks supplying urban areas; by 1826, only two towns in Britain with populations greater than 10,000 lacked a gas company and gas lighting quickly became popular across Western Europe and

Fig 4.5
The lamp room, Castle Coole, has an earlier plunge bath below the floor, which has been boarded over.

Fig 4.6
A paraffin wall lamp on a modified gas lamp bracket, Erddig.

Fig 4.7
A small gasworks sold by
W C Holmes, c 1845. From
A Hundred Years of Service,
W C Holmes.

North America, too.[16] During this period, there was brief interest in manufacturing gas from whale oil, partly to help sustain the whaling industry which had been hit by declining demand for lamp oil, but the process was more expensive and less predictable than using coal, so the idea was not widely adopted. Until the introduction of steam-powered pumps (exhausters) in the late 19th century, gas could only be supplied economically for a distance of about 3km from a typical town gasworks, which meant

that few country houses could take advantage of public supplies. One exception was Powis Castle, located only 1.6km from Welshpool gasworks. Another example was Dunham Massey in Greater Manchester, which was 4km away from the Altrincham Gas Company's works. This would normally place it outside the economic supply area, but in 1868 the Earl of Stamford contributed £150 and also agreed to a premium of around 14 per cent above the normal gas price to pay for extending the mains network with large diameter pipe to Dunham.[17] However, the vast majority of country houses had to build their own gasworks if they wanted to enjoy the benefits of gas lighting, and dozens of small foundries started supplying complete gasworks or components to factories, country houses and public buildings, as well as to the rapidly increasing number of public gas companies. Hatfield House, Hertfordshire, took gas from the local town gasworks in the 1860s but found the supply unsatisfactory and built their own works instead.[18] By around 1850, the market for small gasworks became dominated by a few specialist manufacturers, most notably George Bower of St Neots, W C Holmes of Huddersfield, George Porter of Lincoln and Edmundson & Co of London and Dublin (Fig 4.8). In some cases, the equipment was supplied direct to the house owner by these companies but more often it was purchased by a local engineering contractor who

Fig 4.8
An advertisement for
country house gasworks
from Edmundson & Co,
Dublin and London.
From Walford's County
Families, *1875.*

not only erected the plant but also installed the pipework and the lights.

The lights themselves were mostly naked flames of various shapes (such as batswing, cockspur and fishtail) designed to maximise their brightness, although some larger lamps utilised Argand-type burners. In more polite domestic surroundings, the flames were enclosed in glass shades but in service areas, as in other workplaces, they were generally open. These lights were no brighter than the Argand oil lamps which preceded them but provided light more cheaply and conveniently than oil lamps. Being fixed, gas burners also presented less of a fire hazard than portable oil lamps. Their convenience is illustrated in an account written by Sir Walter Scott's biographer, John Lockhart. Scott was one of the backers of an early oil gas company in Edinburgh and installed a small oil gas plant to light his house, Abbotsford, in Roxburghshire, in 1823. Lockhart described its first demonstration thus: 'Dinner passed off, and the sun went down, and suddenly, at the turning of a screw, the room was filled with a splendour worthy of the palace of Aladdin.'[19] Abbotsford is the first country house known to have been permanently lit by gas; the oil gas plant was replaced by a conventional coal gas one by a subsequent owner.

Gas therefore offered light, if not at the flick of a switch (although such convenience did become available later in the 19th century), then at the turn of a tap, without the labour involved with filling lamps and trimming wicks. Another important advantage was cost. Accounts for Chatsworth House, Derbyshire, show that, in the years preceding the completion of its gasworks in 1862, it spent an average of £202 per year on lamp oil and candles; after 1862, this fell to £46, while the total cost of running the gasworks was £129.[20] Cost savings here and at most other country houses would have been greater had gas lighting been used throughout the house but this was rarely the case. One objection to the use of gas in state rooms was the harshness of the light. Even Lockhart had to temper his enthusiasm: 'The old lamp would have been better in the upshot. Jewelry sparkled, but the cheeks and lips looked cold and wan in this fierce illumination; and the eye was wearied and the brow ached if the sitting was at all protracted.'[21]

Another problem was posed by the acidic fumes produced by the burning gas, which were found to have a deleterious effect on paintings, tapestries, furnishings and book bindings. A few country houses adopted a solution also used in many museums and galleries of installing lights which ventilated the fumes to the outside of the building; a spectacular example is the 'Sun Light' in the Yellow Drawing Room at Wimpole Hall, Cambridgeshire (Fig 4.9). Unusually, this lamp had a control valve, hidden behind a panel in a door case, and a permanent pilot light, because its location made it so difficult to light by conventional means. Most country houses simply

Fig 4.10
A gas light fitting in close
proximity to paintings,
in the Pagoda Room,
Burghley House.

Fig 4.11
The former gasworks,
Welbeck Abbey. The main
building is a reconstruction
of the original retort house;
the smaller building to the
right was the purifier house.
The outlines of two of the
four gas holders can be
seen as landscape features.

restricted their use of gas lighting to service areas and larger, well-ventilated spaces such as entrance and staircase halls. There were some notable exceptions, however: Burghley House, in Lincolnshire, used gas lighting throughout the ground floor rooms, with pipes running exposed in front of the wood panelling and lights close to valuable paintings (Fig 4.10), and the Marble Hall at Hatfield House also had gas lights adjacent to its rare tapestries, which can be seen in Figure 4.20. Small improvements to gas burner design were made through the 19th century but it was only with the invention in 1885 of the incandescent gas mantle – a textile gauze impregnated with rare earth metals which glowed bright white when heated with an aerated Bunsen-type flame – that the illuminating power of gas lights was improved dramatically. These could produce light up to 60 candle power, compared with 10 to 12 candle power for a conventional gas flame, but consumed less gas, so the quantity of fumes emitted was also reduced. By this time, however, gas lighting was facing increasing competition from electricity.

Country house gasworks

Country house gasworks varied greatly in size, depending on the extent to which lighting was used on the estate. One designed in 1877 for Temple Belwood, in Lincolnshire, had only one retort and a gas holder 4.2m in diameter; the cost of the plant, excluding the building that housed it, pipework and lights, was £254.[22] The gasworks at Chatsworth House was more typical, with a 10m-diameter gas holder; the total expenditure on gas lighting for the house, including the gas house, pipework and lights, was around £2,700.[23] Possibly the most extreme example was at Welbeck Abbey, Nottinghamshire, where the gasworks, built in 1869, had four gas holders and was large enough to light a small town (Fig 4.11). The house itself is massive, with many underground rooms, including a library lit by 1,100 gas burners. Gas was also used to light the stables, picture gallery, estate buildings and an extensive network of underground tunnels; the massive Riding School, built the same year as the gasworks, was reportedly lit with around 8,000 gas jets (*see* Chapter 2). Almost 350 examples of country house gasworks have been identified in the course of this research, and there were probably several hundred more in existence by the time gas lighting was supplanted by electricity around the start of the 20th century. Curiously, only a handful of these were built before 1850, most being installed in the period 1860–80. This is in stark contrast to the situation at factories and other industrial premises, where almost 90 per cent of the private gasworks had been built by 1850.[24]

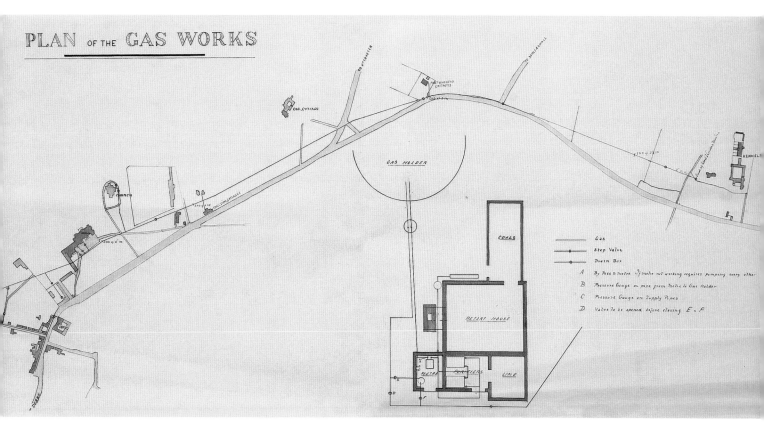

PLAN OF THE GAS WORKS

In many cases, gasworks were located close to the house itself, often alongside the stable block or kitchen garden. The image in the advertisement in Figure 4.8, with the gasworks screened from the mansion by just a belt of trees, is not entirely fanciful (although the inclusion of what are presumably the master and mistress of the house gazing on contentedly perhaps is). At Cliveden, Buckinghamshire, a maze of large shrubs was created to hide the gasworks from the house; the gasworks chimney was also designed, somewhat unconvincingly, to resemble a tree trunk. However, the process of manufacturing coal gas inevitably created smoke and unpleasant smells so many owners chose to site their gasworks some distance from the house, often at the home farm or woodyard. In around 40 examples, the gasworks were located in the nearby village, from where they would usually supply gas to other properties and for street lighting in the village, as well as to the estate itself; examples include Brodsworth Hall in Yorkshire, Cragside in Northumberland and Petworth House in West Sussex. At Sudbury Hall, Derbyshire, a gasworks was built in the village, in an elegant building attributed to George Devey, and gas was supplied to the church, kennels and many other estate properties, as well as to the Hall itself (Fig 4.12). Most estates did this as an act of philanthropy but the gasworks at Dunrobin Castle, Highland, supplied the village of Golspie as a commercial venture and the undertaking became part of the nationalised gas industry in 1948.[25]

The locations of some country house gasworks were designed to facilitate the transport of coal: for example, at Halton in Buckinghamshire and Shipley Hall in Derbyshire they were next to a canal, and at Waddesdon in Buckinghamshire the gasworks was next to the Wooton tramway, from which it had its own siding. Coastal locations were chosen at Castle Ward and Mount Stewart (both in County Down) and at Culzean Castle, in Ayrshire (Fig 4.13) so that coal could be delivered by boat. Substantial remains of the gasworks survive at all three of these sites and at many other houses, too. Indeed, more former gasworks buildings survive on country house estates than anywhere else, as there has been less pressure in these rural locations to demolish these buildings when they became redundant, and many have been adapted for other uses. The below-ground gas holder tanks were often filled with rubble when the works closed, and so survive below the current ground surface.

Fig 4.12
A plan of Sudbury gas house and supply network, 1874. The gasworks building, shown in detail in the centre, is located near the left edge of the plan. It is attributed to the architect George Devey and is still standing; plans are being developed to restore it for community use.

Fig 4.13
The gasworks, Culzean
Castle. Its location close to
the shoreline allowed coal
to be delivered by sea.

Fig 4.14 (facing page top)
The gas house, Myra Castle.
It was originally built as a
coal gasworks, which was
later replaced with an
acetylene plant.

Fig 4.15 (facing page
bottom)
An acetylene plant in an
outbuilding at a private
house in Somerset in 2010.
The plant has since been
removed and is on display
at Hestercombe Gardens.

Many country houses built in the second half of the 19th century had the gas pipework installed as part of the original fabric but for older houses it was necessary to disturb the historic interiors to lay the pipes, perhaps one reason why gas lighting was slow to be accepted in many houses. In state rooms, these pipes were usually concealed (although that was not always the case, as Figure 4.10 illustrates); in service areas, the pipes were generally exposed on the walls and ceilings. Gas supplies rarely extended to sleeping quarters, possibly on safety grounds. Although gas was predominantly provided for lighting, it was sometimes used for cooking, as described in Chapter 6, but very rarely for heating (*see* Chapter 5). One other important use of gas on country house estates was to fuel gas engines. These were the first successful form of internal combustion engine, introduced in the 1870s, before oil, petrol or diesel engines were developed. They provided a convenient alternative to steam or water power for a variety of estate purposes, for example in sawmills and workshops. For a period around the end of the 19th and beginning of the 20th centuries, gas engines were also widely used for electricity generation.

The erection of a gasworks brought a need for new specialist employees. The gasworks for a larger house operated more or less around the clock for most of the year and needed at least one skilled full-time employee, plus occasional additional manual labour. An advertisement placed by the Earl of Shrewsbury's agent at Ingestre Hall, Staffordshire, in 1876 read: 'Wanted, a practical Gas Maker (single-handed), to take charge of and make gas, do gas and water fittings, and if he could do plumbing preferred. A cottage and garden and fuel found.'[26] The Rothschilds, characteristically, used their family networks: the nephew of the gasworks manager at Ascott, Buckinghamshire, was recruited to run the works at Waddesdon, after first spending two years training at Ascott.[27]

In most cases, these private coal gas plants closed when electricity became available, either through private generation or from the mains, and in a few cases the former gasworks buildings were used to house an electricity generator. This was the case at Apley Park, Shropshire, where the retort house was converted to house a steam-powered electricity generator around 1900. There were a few exceptions: Chatsworth's gasworks was still operating in the 1920s, despite the construction of an electricity generating plant in 1893. Burghley House's gasworks only closed in the 1930s when the house came within reach of a public gas supply from Stamford, and the house continued to be lit by gas until electricity was installed by David Cecil when he inherited the house and the title of Marquess of Exeter in 1956.[28] His daughter, Victoria Leatham, was nine years old at the time and later wrote: 'There was a constant hissing from the gas lights and a heavy, sweetish smell pervaded the building. In suspect spots in the corridors leaking gas would make a lit match flare up, threatening the eyebrows and encouraging pyromania among children.'[29] At Cragside, the gas supply was not installed until after the introduction of electric lighting. When the water supply for hydroelectric generation proved to be inadequate, in 1895 Lord Armstrong built a gasworks which provided fuel for a gas engine for electricity generation, but the gas was also used in parts of Cragside mansion and estate, as well as in the village of Rothbury.

Alternative forms of gas lighting

The high cost of building and running a conventional coal gasworks deterred many owners, particularly those of smaller houses, from installing gas lighting. A solution appeared in the form of two alternative types of gas manufacture, developed around the end of the 19th century, which were much cheaper to build and did not require full-time specialist staff. These were acetylene and petrol-air gas. The technique of making acetylene gas from calcium carbide was pioneered in Canada in 1892 and continued in use in devices such bicycle lamps and bird scarers until at least the 1950s. The earliest examples of domestic acetylene plants consisted of separate components and were operated manually. Lumps of solid calcium carbide were placed in one of the reactor vessels, which was then closed. The required amount of water was then added manually from a tank similar to a WC cistern. The gas produced was passed through a purifier and dryer, and stored in a gas holder. This would be sufficient to light a small house for several days, after which the process would be repeated. An intact example of such a plant survives at Myra Castle, County Down, where the present owner recalls that his mother-in-law, as a child, regarded it a treat to be allowed to visit the gas house on a

Saturday morning to help make the gas (Fig 4.14).[30] Another complete example of a manual plant, supplied in the 1890s by the St James' Illuminating Company of Victoria St, London, was removed from the grounds of a private house in Somerset in 2010 and is now displayed at Hestercombe Gardens, Somerset (Fig 4.15).

In the early 20th century, self-contained plants were developed which were more automated in their operation: as the gas holder

Fig 4.16
*An oil ceiling light
converted to acetylene gas,
The Argory.*

Petrol-air gas plants had a lower capital cost than acetylene – sometimes as little as £20 – and required even less labour to operate them. The earliest description of such a plant dates back to 1872 and a 1909 publication describes about ten different designs of petrol-air plant, all of which operated on the same principles.[31] Petrol was pumped from a storage tank to a carburettor, where it was mixed with air introduced by a positive displacement blower. The resultant flammable mixture was stored in a small reservoir. The ratio of fuel to air had to be maintained within narrow tolerances if the mixture was to burn correctly and not explode, but it could be burned in aerated burners similar to those used for coal gas, so this gas, unlike acetylene, could be used with incandescent mantles, gas cookers, fires and other appliances. Among the many manufacturers of these plants was Edmundson of London and Dublin, who also supplied conventional coal gas plants and, later, electric lighting installations for country houses. In the 1880s they supplied two plants to St Michael's Mount, Cornwall, one for lighting and the other to power a gas engine which drove the tramway (*see* Chapter 8).[32] Petrol-air gas plants were fully automatic in operation and the main variation between the different designs was the source of power to drive the fuel and air pumps. Most designs utilised weights on pulleys (Fig 4.17), which needed to be wound up every few

emptied, water was automatically introduced into the reactor vessels to produce more gas. One example, manufactured by the Acetylene Corporation of Victoria St, London, remains in situ at The Argory, County Armagh. Acetylene cannot be used with the aerated, Bunsen-type burners needed for incandescent mantles, so this form of gas lighting had to rely on luminous naked flames. Usually, a pair of burners was mounted in such a way that their flames impinged, which gave a slightly brighter light (Fig 4.16). Acetylene was not suitable for cooking or heating so its use was confined to lighting.

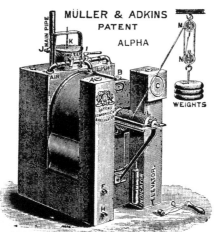
Fig 4.17
*An advertisement for
Muller's petrol-air gas
generator. From* The
Country Gentlemen's
Catalogue, *1894.*

days; an example, made by the Aerogen Gas Company of London, survives in the stable block at Rufford Old Hall, Lancashire. Other designs used hot air, clockwork or compressed air motors or water turbines to pump the petrol and the air.

While the main application of acetylene and petrol-air gas plants was to provide gas lighting for houses that had previously had none, they were occasionally used to replace earlier coal gas plants which had presumably been considered too expensive to maintain: the acetylene plant at Myra Castle occupies the retort house of a former coal gasworks (see Fig 4.14). The coal gas plant at Holkham Hall, Norfolk, built in 1865, was replaced with three 'Loco Vapour Gas' petrol-air gas plants in 1908; the results were evidently not satisfactory and an electricity generating plant was installed the following year.[33] Petrol-air gas was unlikely to prove effective for large properties as temperature variations in an extensive pipe network would have affected the vapour concentration and could even have caused petrol to condense out.

Electricity

Early experiments in electricity from the late 18th century and through most of the 19th century used chemical batteries (primary cells) to provide the power; the most successful later types being the Daniel and the Leclanché cells.[34] As with gas, most early demonstrations of electrical power were little more than novelties, but batteries such as these provided the power for the first practical application of electricity: for telegraph systems in the mid-19th century, for electric bell systems (see Chapter 7) and (with limited use) for early lighting systems from the 1870s. In the 1830s, Michael Faraday and others showed how an electric current could be created by moving a conductor through a magnetic field, which led to the development of the dynamo in the 1850s, enabling electricity to be generated from mechanical power; the invention of the lead-acid (Planté) secondary cell in 1859 then allowed the electrical power produced by a dynamo to be stored, providing an alternative to primary cells.[35]

The first form of electric lighting was the arc light, in which illumination came from a continuous spark created between two, usually carbon electrodes. The light produced was very bright and harsh, and the lamps were noisy and produced smoke, so they were mostly used out of doors. Arc lamps, often powered by portable steam-powered generators, began to be used to illuminate outdoor events during the 1870s, including the world's first floodlit football match, which took place in Sheffield in October 1878 (Fig 4.18), and for early trials of electric street

Fig 4.18
The world's first floodlit football match, Bramall Lane, Sheffield, 14 October 1878. From Illustrated Sporting and Dramatic News, 16 November 1878.

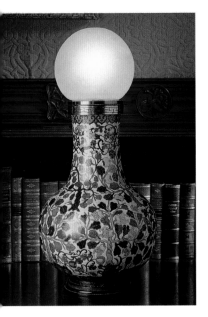

Fig 4.19
A cloisonné enamel vase
converted to an electric
lamp in 1880, Cragside.

lighting.[36] An account of the County Ball at Hatfield House in January 1874 states:

> Two powerful arc lamps were placed in a type of sentry box – made for that purpose – near the centre flower bed on each side of the South Front, and another on the wall at the east side of the North Front, and by means of batteries a good light was kept through the night, a great boon to the enormous amount of traffic on such an occasion.[37]

Despite the obvious drawbacks of arc lights, some pioneers of electric lighting experimented with their use indoors: Sir David L Salomons in his workshops and elsewhere at Broomhill, Kent, in 1874, Lord Armstrong in the Picture Gallery and workshops at Cragside in 1878, and Lord Salisbury in the Marble Hall and dining room at Hatfield House in 1880.[38]

A number of inventors worked through the 1870s to find a form of electric light that was more suitable for indoor use and, although Thomas Edison often receives the credit for this, most historians accept that it was the Newcastle-based chemist Joseph Swan (who later went into partnership with Edison) who produced the world's first successful incandescent electric lamp in 1878, which he patented in 1880.[39]

Sir William Armstrong presided over one of the first public demonstrations of Swan's lamps and quickly installed them at Cragside, fed with electricity by the water-powered dynamo that had previously supplied his arc lamps. In a letter to *The Engineer* in January 1881, Armstrong provided a detailed description of this installation, which consisted of 45 electric lamps (soon increased to 92).[40] Some of these were mounted on cloisonné vases which had previously held kerosene lamps; the electric current flowed through the metal bodies of the vases and the mercury-filled cups in which they stood, so that the lamps could be extinguished by removing them from the cups. Two of these lamps survive in the library at Cragside, although they no longer have electric current flowing through the bodies (Fig 4.19). A few days after Armstrong's letter appeared, the 3rd Marquess of Salisbury wrote to Armstrong, seeking his advice about lighting at Hatfield House. Armstrong replied recommending the services of Siemens Brothers, but also suggested seeking help from Swan himself, describing him as 'a good electrician', who 'ought to know best what will suit his own lamps'.[41] Lord Salisbury obtained a supply of Swan's bulbs, at a cost of 25s each, which he suspended from the ceiling of the Marble Hall (Fig 4.20).[42] A visitor to Hatfield later in the 1880s,

Fig 4.20
Marble Hall, Hatfield House in the early 1880s. A number of Joseph Swan's incandescent electric bulbs can be seen hanging from the ceiling. The picture also shows the earlier gas lamps in close proximity to valuable 17th-century tapestries.

when the lighting system had probably been extended, wrote: 'You could only just see the red colouring on the majestic old house, but all the windows blazed and glittered with light through the dark walls; the Golden Gallery with its hundreds of electric lamps was like a Venetian illumination.'[43]

Sources of electrical power

The Swan lamps at Hatfield were initially powered by batteries – work to convert a water-powered sawmill on the estate for electricity generation was not completed until the following year. Batteries, in the form of either primary cells or rechargeable secondary cells, played an important part in gaining acceptance for electricity for domestic lighting among rich and influential patrons. The Electrical Power Storage Company and others provided temporary lighting installations for balls and other grand occasions, necessitating the transport of large numbers of glass-case lead-acid batteries around the country.[44] In June 1883, a State Ball at Buckingham Palace was lit with a temporary installation of 700 Swan lamps by W H Massey, with which the Prince of Wales was reportedly 'very pleased'.[45] However, this form of lighting needed a more permanent electricity supply than batteries if it was to flourish. Public electricity networks began to develop in London and a few other major urban centres during the 1880s but the range of these systems – invariably operating at 110 volts DC – was even more limited than that of early gas networks, so wealthy 'early adopters' wishing to take advantage of this exciting new development often had to install their own generator plants in their town houses as well as at their country estates.

For estate owners with access to water power, this was generally the preferred means of providing electricity from the earliest days of electric lighting until the arrival of mains power, because of its minimal running costs. While a few early experiments used traditional waterwheels, including the world's first electric street lighting, installed in Godalming in 1881, most country house hydroelectric plants adopted the more compact and efficient vortex turbine, which had been developed by James Thompson in Belfast around 1850.[46] The leading manufacturer of this turbine was Gilbert Gilkes & Co of Kendal, who are still operating. Some turbines were installed either in or alongside the wheel pit

of watermills, to take advantage of the pre-existing water supply infrastructure; examples include Hatfield House (1882), Alnwick Castle in Northumberland (1889), Warwick Castle (*see* Fig 1.18) (1894), Hestercombe, where a sawmill was extended by Lutyens to house the generator plant (1895), and Bateman's in East Sussex (1902) (Fig 4.21). At Alnwick, the waterwheel was still being used to drive the corn mill, which limited the water supply to the turbine, and attempts were made to use a redundant steam traction engine to supplement the power for electricity generation, before an oil engine was installed in 1895.[47] Many more country houses installed hydroelectric generation in purpose-built turbine houses. Perhaps the earliest example, the *c* 1878 turbine house at Cragside, was replaced with a new one in 1886. This was closer to the house, with a new water supply system to provide a greater head. (As mentioned earlier in this chapter, the turbine was later supplemented by a gas engine.) At Chatsworth, the system that provided water at a 120m head of pressure to the Emperor Fountain was adapted in 1893 to feed a new turbine house, discreetly built into the garden terracing (*see* Chapter 2 and Fig 10.11). Other estates at which there are substantial remains of the hydroelectric generation plant include Knightshayes Court

Fig 4.21
A water turbine for electricity generation, installed in 1902 in a waterwheel pit, Bateman's.

in Devon (*c* 1900), Ardkinglas in Argyll (1908) and Castle Drogo in Devon (1911). Chapter 10 describes how a number of these early water turbine sites are being developed for modern power generation.

While estates in more hilly regions could take advantage of cheap water power, those in lowland Britain had to use other forms of power. Initially, the only alternative was the steam engine, which was expensive both to install and to run. Among the earliest country houses to install a steam generation plant was Berechurch Hall in Essex (now demolished), which was totally rebuilt for the MP and brewer Octavius Coope in 1882, and was thus perhaps the first country house to have been constructed with electricity from the outset. An article in *The Electrician* in 1882 describes and illustrates the installation: it had one Davey-Paxman 'under-type' engine (with the boiler and engine mounted on a single frame) driving four dynamos, which lit 200 Swan lamps.[48] The plant was designed by, and installed under the supervision of, Colonel Rookes Crompton, who had recently set up a business manufacturing dynamos in nearby Chelmsford, which later developed into a major electrical engineering company. 'Undertype' or locomotive-style steam engines were popular for country house electricity generation because the integrated boiler and engine saved space, but many installations (especially larger ones) used more conventional horizontal engines with a separate boiler room, for example at Lyme Park in Cheshire (Fig 4.22), Callaly Castle, Northumberland, and Ascott, Buckinghamshire, where the elegant 1890s generator house complex still contains its large Cornish boiler, supplied by Marshalls of Gainsborough. As well as having a high capital cost, steam engines needed constant attention from a skilled operator, so were generally restricted to larger country houses. The smoke and steam they produced could also be obtrusive: at Wightwick Manor, strictly a suburban villa rather than a country house, built in 1887 near Wolverhampton, the steam-powered electricity generator, housed in an outbuilding, provoked complaints from neighbours about the smoke.[49]

Gas engines, as described earlier in this chapter, provided a lower-cost alternative to steam if a gas supply was available. The earliest known example at a country house was at Didlington Hall in Norfolk (now demolished), where Edmundson & Co installed a generator

Fig 4.22
The 1904 steam-powered generator plant, Lyme Park.

driven by a gas engine in 1882, supplied from the gasworks they had built 11 years earlier.[50] Gas engines became more popular after around 1900, with the development of the suction producer gas plant, which produced gas much more cheaply and easily than a conventional gasworks (Fig 4.23). In these, coal or other organic material was heated in a closed iron vessel and air was sucked through it by the engine, producing a low-quality gas which was not suitable for use for lights or other appliances but could power an engine. At Holkham Hall a pair of gas engines and dynamos, each with its own suction producer gas plant, was installed in 1909; in normal operation, these were only used alternate days and the plant was designed to supply over 1,900 lamps.[51] Also around 1900, hot-bulb oil engines, which could run on a range of different liquid fuels, became sufficiently powerful and reliable to compete with gas engines and were soon the preferred type of prime mover for generating electricity (and many other applications) on country house estates, if water power was not available. During the course of this research, almost 500 examples of country house generator plants have been identified, of which 11 per cent used water power, 14 per cent gas engines, 34 per cent had oil engines and 41 per cent used steam power (including three estates which had steam turbines). What is most remarkable about this analysis is the speed at which electric lighting was embraced: despite the considerable cost of providing a private electricity supply (which is explored further later in this chapter), over 96 per cent of this sample had done so by 1905, within around 25 years of the first use of incandescent bulbs. A large number of new companies sprang up in the last two decades of the 19th century to satisfy the growing demand for electric lighting equipment and some of these focused particularly on the country house market, most notably Drake and Gorham, whose founder, Bernard Drake, worked on some early experiments at Hatfield House, and who had installed electric lighting in over 300 country houses by 1905, including Chatsworth House and Cragside.[52] Other notable early installers included Edmundson & Co, who had previously supplied many country houses in Britain and Ireland with gasworks, and W H Massey, whose reputation among aristocratic clients must have been enhanced by his appointment as 'Mechanical and Electrical Light Engineer' to Queen Victoria in 1885.[53]

Almost all of these installations operated at 110v DC, although a few, built after 1900, are

"CAMPBELL" SUCTION GAS PLANTS.

Fig 4.23
An advertisement for 'Campbell' suction producer gas plants. From the 1910 catalogue, Campbell & Co, Halifax.

known to have used 230v DC. Lord Salisbury experimented with a higher voltage DC supply at Hatfield House in 1882, using a dynamo designed for use with arc lights supplied by the American Brush Company, but the trial was abandoned after it resulted in the death by electrocution of one of the estate's gardeners.[54] No evidence has been found for AC generation at a country house, although a small private house in Cumbria is known to have used it from 1896.[55] A major disadvantage of 110v supplies was that the high current caused significant losses in the cables, which meant the generator plant needed to be close to the house, usually no more than 1km away. However, the advantage of DC systems was that power could be stored in batteries. The large lead-acid Planté-type cells each produced only 2v, so a bank of 55 of these was needed for a 110v domestic supply (and 115 for a 230v supply). The batteries were housed in a room that was physically separated (and occasionally remote) from the generator house, with good ventilation to prevent the accumulation of the hydrogen given off during the charging process.

One of the pioneers of this form of electrical storage was Sir David L Salomons who was, as described earlier in this chapter, perhaps the first person to light parts of his house with arc lamps. Salomons inherited the Broomhill estate from his uncle, a successful banker, in 1873, while still a student at Cambridge. The nephew showed no interest in the family business but instead devoted his energies to scientific experimentation and new technologies. He extended his relatively modest country house with a large, galleried laboratory designed for public demonstrations, and an impressive motor house (*see* Chapter 2 and Fig 2.15). Salomons wrote numerous books and articles on electricity and motoring, one of which, *Electric Light Installations and the Management of Accumulators*, included details of his own battery room, which served as a model for many others that were subsequently built (Fig 4.24).[56]

The use of batteries provided continuity of supply in the event of a short-term failure of the generator plant but also meant that the plant did not have to be run continuously when electricity was required. Detailed logs of the operation of the generators at Holkham Hall show that the plant was typically run for three to four hours during the day in winter to charge the batteries, which then lit the house in the evening and overnight; the longest running time was seven hours, on Christmas Day 1910.[57] Most country houses

probably followed a similar pattern, which avoided the need for the staff operating the plant to work late into the evening. Some country house internal telephone systems (*see* Chapter 7) included an extension in the electrician's house, so he could be called to restart the plant should dimming of the lights show that the batteries were beginning to run down.

Using electricity

Electricity, like gas, was introduced predominantly to provide lighting but, unlike gas, other applications existed from the outset, notably for bells and telephones (although, as explained in Chapter 7, these were often initially powered by batteries) and more uses were developed rapidly. Some early applications of electricity, like electric cigar lighters and dinner gongs, were little more than novelties, showing off the capabilities of this wonderful new phenomenon

Fig 4.24
The Battery Room,
Broomhill, c 1890.

Fig 4.25 (far left)
An Argand oil lamp
converted to electricity,
Yellow Drawing Room,
Tatton Park.

Fig 4.26
An early 20th-century
electric light fitting, visitors'
staircase, Wightwick Manor.

rather than significantly improving the comfort or efficiency of the household, but electric cooking and heating appliances quickly found favour in some country houses, as described in chapters 6 and 5 respectively.

Individually, early electric lights were not necessarily any brighter than gas lamps, especially after the invention of the incandescent mantle had dramatically improved the power of gas lights, but they produced less heat, emitted no fumes and consumed no oxygen from the room, so could be used in large numbers to produce unprecedented levels of illumination. The development of smaller bulbs provided greater flexibility in the design and positioning of light fittings, leading to a growing recognition of the decorative potential of lighting. Chandeliers were sent away for conversion to electricity, using small candle-shaped bulbs, or were replaced with new 'electroliers' (which generally had the wiring more neatly concealed than the converted examples). Many ceiling- and pedestal-mounted Argand oil lamps were converted to electricity, although identification of these is often complicated by the fact that this style was popular for new electric lights, so close examination is required to distinguish between genuine Argand oil lamps that have been converted to electricity and later reproductions (Fig 4.25; see also Fig 5.15). Before long, however, more modern designs of light fittings emerged, befitting this new form of energy, rather than mimicking the styles of previous light sources (Fig 4.26).

Descriptions of the effects which could be achieved with glass and fabric lampshades were illustrated in an influential 1891 book, *Decorative Electricity*, written by Mrs J E H Gordon but with significant input from her husband, who was an electrical engineer.[58]

Early wiring systems were very crude and potentially hazardous. Lord Salisbury's grandson reported an incident at Hatfield House: 'One evening, a party of guests entering the Long Gallery found the carved panelling near the ceiling bursting into flames under the contact of an overheated wire ... and with well-directed volleys of sofa cushions, rendered the summoning of a fire engine unnecessary.'[59] Insulation, using a variety of materials including bitumen, rubber and oil-impregnated fabric or paper, was easily damaged and most early systems kept the live and neutral conductors apart, often running in specially designed wooden trunking, with separate channels for each wire. Fuses, with the fusible links held in small circular holders, were often distributed all over the house. Increasingly robust standards were introduced by the Society of Telegraph Engineers and Electricians (subsequently the Institution of Electrical Engineers) but these were voluntary and it was often insurers who imposed the higher standards. Some country houses, including Tatton Park, in Cheshire, and Stokesay Court, in Shropshire, show evidence of the original 1880s distributed fuse systems having been replaced with the more familiar centralised fuse boxes with

Fig 4.27
Early 20th-century fuses
and circuit-breakers,
Tatton Park.

Fig 4.28
'Jelly mould' switches
(1884), concealed behind
a hinged door, at
Halton House.

ceramic fuse holders (Fig 4.27). At Stokesay, correspondence shows that this was carried out on the insistence of the insurers, to the chagrin of the owner, Herbert Derby Allcroft, who held the original installers, Edmundson & Co, partly responsible for this.[60]

The most popular form of early light switch was the 'jelly mould' design, usually with a polished metal cover; often these were visible on the wall of the room but in some cases they were hidden from view (Fig 4.28). Many original

examples survive but modern reproductions have also been installed in some properties. Manufacturers soon offered a wide range of more elegant switches for use in higher status rooms. The earliest electrical sockets were small, with only two round pins. They were designed for use mostly for table lamps, although most other smaller electrical appliances were plugged into these or into lamp sockets, and hence had no earth connection, until three-pin sockets were introduced in the 20th century. Only larger appliances like electric cookers had more substantial fixed electrical connections.

As with gas lighting, the arrival of electricity usually brought a new breed of employee onto the estate, to operate the generator plant and to maintain the lights and wiring; the rapid growth in popularity of this form of lighting meant that the requisite skills were often in short supply. Hydroelectric plants were the simplest to operate, so it was sometimes possible for the installer to train existing estate staff to operate these – at Hatfield House, this role fell to the Clerk of Works, Henry Shillito.[61] Many marine engineers had the skills to look after steam-powered generators and one such, Walter Thomas, was recruited by Nathan Rothschild to run his plant at Tring Park, Hertfordshire, and to supervise the construction of the hydroelectric plant at the family's Northamptonshire estate at Ashton Wold.

The vast majority of the population was initially baffled by this strange new form of energy which must have added to the status of the electrician. The poet Hilaire Belloc, whose 1893 work celebrating Oxford's new power station includes the lines: 'Awake, my Muse! Portray the pleasing sight that meets us where they make Electric Light,' subsequently wrote the following cautionary ode:

> Lord Finchley tried to mend the Electric
>   Light
> Himself. It struck him dead: And serve him
>   right!
> It is the business of the wealthy man
> To give employment to the artisan.[62]

Blenheim Palace, in Oxfordshire, where the 8th Duke of Marlborough conducted his own experiments with electricity and corresponded with Thomas Edison in America, had four highly valued electricians on its staff; Luton Hoo, in Bedfordshire, reportedly employed ten.[63]

The capital cost of private electricity plants varied considerably: Sir David L Salomons spent

£6,000 on the steam-powered plant which lit his house, Broomhill, from around 1883, while smaller hydroelectric and gas engine installations could be built for between £500 and £1,000.[64] At Chatsworth, labour costs for operating the hydroelectric generator plant averaged around £250 a year (on top of the cost of the gasworks, which continued in use after electricity was installed) and equivalent plants powered by steam, gas and oil engine plants would generally have cost more because of the need to pay for fuel.

Mains electricity reached most rural areas by the 1930s, although some parts of Britain were not connected to the national grid until after the Second World War. Its arrival meant that these private generation plants could be shut down but raised a new problem: the 110v DC wiring, lighting and other appliances that most country houses had installed were generally not suitable for the new 220–240v AC supply. Owners were faced with the decision of either replacing their wiring and appliances or converting the incoming supply to 110v DC. Most chose the former option but in 1933 Holkham Hall installed a converter plant, including a 1.8m-tall mercury-filled rectifier, which allowed rewiring to be deferred until the 1950s.[65] The arrival of the mains supply at Waddesdon in 1926 allowed electricity to be extended to the parts of the estate which had not previously enjoyed this benefit, but the mansion itself remained with a 110v DC supply, provided by a pair of rotary converters (DC dynamos driven by AC motors) which charged the batteries, as the earlier engine-driven generator had done; the massive panel that controlled this equipment survives in the Power House (Fig 4.29).

In conclusion

It is evident, therefore, that developments in lighting, for those able to afford them, brought significant improvements in the comfort and efficiency of the country house from the late 18th century onwards. Whether they reduced the overall workload of servants is less clear: the saving in the time the indoor staff previously spent attending to oil lamps and candles has to be offset against at least one full-time employee (and sometimes more) needed to run a conventional coal gasworks or electricity generating plant.

Often, house owners' investment in new lighting technology was quite short-lived: when large numbers of country houses did eventually

Fig 4.29
The control panel for the incoming AC mains supply and for the plant that converted it to DC in the Power House, Waddesdon Manor.

invest in gas lighting, from the 1860s onwards, they found this technology overtaken by electricity within as little as 20 years, so houses which had been disrupted for the installation of gas pipes were soon being disturbed again for electrical wiring. Acetylene and petrol-air gas lighting often had an even shorter life: for example, Gordon Castle, in Moray, installed an acetylene lighting plant in 1901, which was replaced with a hydroelectric system in 1906. One of the claimed advantages of these alternative forms of gas lighting was that the plant did not need specialist staff to operate it. In 1911, the engineer Maurice Hird wrote: 'The amount of attention required to operate an air-gas or acetylene plant is very small, and it is not unusual to make it part of the housemaid's duty. In the case of electric light the writer has never heard of this having been done; male attention of some sort appears to be required.'[66] Hird went on to suggest that suction producer gas plants could be looked after by a gardener but conventional coal gas plants and steam-powered electricity generator stations needed experienced employees to operate them. These were normally recruited from outside the enclosed world of estate staff, although it is clear that some owners sought to avoid this by having existing staff trained in the new skills required.

Despite the potential cost savings which gas lighting could provide, and its popularity in other areas, this form of lighting was slow to be adopted by country houses. This is in stark contrast to electric lighting, which was embraced with great enthusiasm by many country house owners, despite the increase in household costs it brought. The reasons for this difference are not entirely clear: there is no doubt that electricity was cleaner and more versatile than gas, but its instant success may have been partly due to fashion. Changes in the nature of country estate ownership in the 60 years between the development of gas and electric lighting may have played a part in this: industrialists like William Armstrong and financiers like Sir David L Salomons made important contributions to the development of electrical technology, but this new form of energy also excited the interest of established aristocratic figures like Lord Salisbury and the Duke of Marlborough.

5

Heating and ventilation

'The system of warming our rooms by the air we breathe is not natural.'

J J Stevenson, 1880[1]

Along with lighting, heating can be considered one of the fundamental requirements for domestic comfort, and yet it was an area in which country house owners were often slow to innovate. The open fireplace, burning wood or coal, continued to provide the main source of country house heating well into the 20th century. This was due in part to technical constraints, which limited the effectiveness of early central heating systems. More potent, however, was an awareness, often bordering on an obsession, of the importance of ventilation, which arose from a greater understanding of human physiology. The eminent Scottish physician Andrew Combe wrote in 1834: 'There is scarcely a day passes in which a well-employed medical man does not meet with some instance in which health has suffered, or recovery been retarded, by the thoughtless and ignorant disregard of the value of pure air to the well-being of the animal economy.'[2] Such strong assertions influenced the thinking of country house architects and owners throughout the 19th century and into the 20th.

Open fires

In his influential 1880 book *House Architecture*, J J Stevenson wrote:

For heating English houses, the best system, on the whole, is the old one of open fires. No doubt it is unscientific: it produces dust which by other systems might be avoided; and it is wasteful, as a fraction of the fuel in a close fire would produce the same heat in the room which in an open fire goes mostly up the chimney. But it has the advantage that we are used to it, and that one understands it:

it is a tolerably efficient mode of ventilation, and it is so cheery and pleasant that we are not likely to abandon it, as long as the coal lasts.[3]

The medieval hall generally had a central hearth, a good example of which survives in the Great Hall at Penshurst Place, Kent; these central fires were important sources of light as well as heat. The concept of the central hearth was revived, a little idiosyncratically, by the architect William Burges in the 1870s for the medieval-style Banqueting Hall at Castell Coch, Cardiff, although this was mainly decorative, as the house had an extensive central heating system. Fireplaces with chimneys built into the walls of rooms were introduced to England from France around the late 11th century but it took several hundred years for these to become universal in country houses. This development freed the centre of the hall for use and became essential as the quest for

*Fig 5.1
A fireplace for burning logs, Ightham Mote Great Hall, originating from the 14th century.*

privacy prompted the creation of multiple separate rooms for use by the higher-status members of the household. After this, limited further evolution of the open fireplace took place, with subsequent developments being as much aesthetic as functional.[4]

From the 17th century, coal increasingly became the preferred fuel in many domestic and other locations but the abundant supply of timber on their estates meant that country houses often continued to rely heavily on wood for their open fires. Hearths designed primarily for burning logs tended to be large and confined to halls and other principal ground floor rooms, with the fire contained in raised dog grates (Fig 5.1). Such designs became popular again during the 19th-century Gothic revival era, along with the 'inglenook', which created a small seating area that made the most of the radiated heat from the fire; examples of these can be found in many houses, including Wightwick Manor, in the West Midlands, and Cragside, Northumberland. Large open fireplaces were hugely inefficient and, in all but the smallest rooms of country houses, they still left much of the space uncomfortably cold in winter, despite the frequent use of movable screens to contain the heat. The problem was eloquently described in the memoirs of Barbara Charlton, who arrived at her new home, a country house in a remote corner of Northumberland, as a newlywed in the 1840s:

> Coal and firewood they had in great abundance, it is true, but the long passages had no heat, the outside doors were never shut, the hall and corridors were paved with flagstones while, to complete the resemblance of Hesleyside to a refrigerator, the ... large, old-fashioned full-length windows halfway up [the staircase], with their frames warped by the excessive damp ... contrived to make the downstairs space a cave of icy blasts.[5]

Improved fireplaces

The American Benjamin Franklin was one of the first to apply scientific thought to the inefficiency of open fires. Writing in 1745, he claimed that at least 83 per cent of the heat from these was wasted, adding that:

> Their inconveniences are, that they almost always smoke if the door be not left open; that they require a large funnel and a large funnel carries off a large quantity of air ... and the cold air so nips the backs and heels of those that sit before the fire that they have no comfort till either settles or screens are provided (at considerable expence) to keep it off, which both cumber the room and darken the fire-side ... In short, it is next to impossible to warm a room with such a fire-place: and I suppose our ancestors never thought of warming rooms to sit in; all they purposed was to have a place to make a fire in, by which they might warm themselves when cold.[6]

His solution, which he called the 'Pennsylvanian Fireplace', was a much smaller, recessed fireplace, designed to burn coal rather than logs, and was widely copied. From the 1780s, a variety of different designs of fires with cast-iron fronts emerged, differing only in their aesthetics, bearing names such as Bath, Pantheon and Forest stoves. In the 1790s, the American-born scientist Benjamin Thompson, Count Rumford, demonstrated a more scientific approach to the design of chimneys to improve their efficiency and developed shallower fireplaces which radiated more heat into the room (Fig 5.2). Rumford's ideas borrowed heavily from the work of Franklin and others but his designs were still being recommended three-quarters of a century later, a testament, perhaps, not just to their effective-

Fig 5.2
'The Comforts of a Rumford Stove', etching by James Gilray, c 1800.

ness but also to Rumford's notorious abilities at self-promotion. His designs found their way into new country houses as well as lesser dwellings.[7]

Another enthusiastic promoter of improved fireplace design was the radical writer William Cobbett, who was no doubt introduced to these developments during his time in America in the 1790s. Cobbett wrote about and illustrated 'American fireplaces' in his weekly newspaper, the *Political Register*, in the 1820s, and even displayed examples he had imported in the newspaper's offices in London.[8] The design was quickly copied by British manufacturers, including a Mr Judson of Kensington. Confusingly, a development of these fireplaces, with cast-iron sides and back as well as front, was known as the Register stove, but the name probably derives from the adjustable plate or register, fitted at the mouth of the flue to regulate the flow of flue gases and to control the ventilation provided by the chimney when the fire was not lit.

Attempts were sometimes made to distribute the heat from open fires more widely through the house, for example by running the flues from downstairs fireplaces up the inside of the walls of the rooms above. At the back of most of the fireplaces on the ground floor at Wollaton Hall, Nottingham, there are iron tubes which it is believed were designed to heat air which was ducted to the upper storeys of the building; these probably date from the early 19th-century modernisation undertaken by Jeffry Wyatville (Fig 5.3). Lewis Wyatt, in his work at Tatton Park, Cheshire, in the early 19th century, adopted a slightly different approach to the challenge of distributing the heat from open fires more widely: the backs of the brick or stone fire-backs of the two fireplaces in the entrance hall were exposed behind grilles in the wall of the staircase hall behind.[9] At Holkham Hall, Norfolk, work to install a new fireplace in a room under the Statue Gallery in 1757 included 'turning the flew of the works around the cavity to heat the above Gallery'.[10]

Flues were often linked together and access hatches were provided to facilitate cleaning. Chimneys were relatively expensive to construct and, particularly from the 16th century, a multiplicity of them was a conspicuous demonstration of wealth. Chimney pots were sometimes disguised or hidden behind parapets to avoid disfiguring the roof lines of classical houses but often became design features in their own right in the more flamboyant 19th-century architectural styles. In typically dogmatic fashion, the architect J C Loudon advised, 'Every person that has a house designed for him ought to object to every chimney top, whether Grecian or Gothic, that does not consist of an obvious base, shaft and capital.'[11]

The growing size and complexity of country houses from around the 17th century onwards increased the number of fireplaces in each house, which created a massive workload in clearing, making and maintaining fires. Transportation of fuel within the house also required a huge amount of servant effort and Chapter 8 describes methods adopted in some houses to minimise the disruption this would otherwise have caused. However, delivery of coal to individual rooms could never have been unobtrusive, as this recollection by Lady Diana Cooper of life at Belvoir Castle, Leicestershire, around the start of the 20th century shows:

If anyone had the nerve to lie abed until eleven o'clock, which can seldom have happened, there were many strange callers at the door. First, the housemaid, scouring the steel grate and encouraging the fire from the night before, which always burned until morning, and refilling the kettle on the hob until it sang again. Next the unearthly water giants then a muffled knock given by a knee, as the coalman's hands were much too dirty and too full. He was a sinister man, much like his brothers of the water, but blacker and far more generally mineral. He growled a single word, 'Coal-man' and refilled one's bin with pieces the size of ice-blocks.[12]

Fig 5.3
A fireplace in the Great Hall at Wollaton Hall, with metal tubes at the rear of the grate to heat air for the floor above.

Despite its practical limitations, the open fireplace has remained the physical and decorative focus of most rooms in country houses: artworks and furniture are arranged around it and bell handles for summoning servants (*see* Chapter 7) are almost invariably located either side.

Enclosed free-standing stoves

Enclosed stoves offered significant practical advantages over open fireplaces, not least because the control of air supply meant that fuel burned more slowly. More importantly, radiation and convection from the body of the stove and any exposed flue transferred a greater proportion of the heat into the room. Such stoves eventually became common in Britain in the 19th century, particularly in churches and other

public buildings, with many early designs betraying American or European origins, building on work by pioneers such as Franklin. One intriguing character who helped introduce this type of heating apparatus from Europe was Abraham Buzaglo, the son of a Moroccan rabbi, who arrived in London around 1762 and was granted a patent for a multi-tiered free-standing cast-iron stove in 1765.[13] The only known surviving example of his products in Britain is at Knole, in Kent, which is marked with the date of 1774 (Fig 5.4). This now stands in the Orangery, where it was moved for storage in the 19th century, but was originally installed in the Great Hall, in front of the present fireplace, with a flue pipe running from the back of the top tier of the stove into the wall above the mantelpiece. One other example survives in Williamsburg, Virginia. Buzaglo later promoted his stove as a cure

Fig 5.4
A Buzaglo stove, originally installed in the Great Hall at Knole but now in the Orangery.

for gout and died in Soho, London, at the age of 72 in 1788.[14]

More British and European inventors moved into this field, including the Marquis de Chabannes in 1818, and the Scottish doctor Neil Arnott, physician extraordinary to the Queen, who published details of what became known as 'Arnott's Stove' in 1838.[15] A stove similar to his design survives in the Grooms' Room at Wollaton Hall. In the 19th century, various designs of enclosed stove continued to be more popular in many parts of Europe than in Britain, particularly for domestic use – one 1845 book about heating describes what were popularly known as German, Dutch, Russian and Swedish stoves.[16] Few examples now survive in country houses in Britain, and one obvious reason for this was aesthetic: these large cast-iron stoves detracted from the elegance of a polite interior. A notable exception is to be found at Castle Coole, County Fermanagh, where the inner lobby on the first floor contains a pair of small elegant iron stoves, standing at dado height in curved recesses (Fig 5.5). James Wyatt designed the house and much of its furniture around 1790 and drawings show that he was also responsible for the design of these stoves. In the interests of symmetry, he installed two identical dummy stoves in recesses in the opposite wall. The flues for the two working stoves are shared with a pair of larger cast-iron stoves located in curved recesses in the Saloon below. These are of a more ornate design, topped with urns, and may have been part of the original fabric of the house or could be of a slightly later date. James Wyatt was evidently an early adopter of free-standing stoves; his first major work, The Pantheon in London, which opened in 1772, used four large cast-iron stoves in niches to heat the Great Hall.[17] The Saloon at Kedleston, Derbyshire, also contains a pair of cast-iron stoves, together with a pair of dummy stoves, located in niches, believed to have been designed by Wyatt in the 1770s. In store at Weston Park, Staffordshire, are the remains of a cast-iron stove made by the Carron Company in Falkirk around 1780, to a design by Robert Adam.

An equally elaborate free-standing stove, again topped with an urn, can be found at The Argory, County Armagh; an architect's drawing suggests this stove dates from around 1821. It is located in the centre of the West Hall, at the foot of the main staircase, which would have allowed some heat to rise to the first floor. This evidently proved ineffective and a second stove was installed immediately behind the first. This

Fig 5.5
A stove designed by James Wyatt, c 1790, first-floor lobby, Castle Coole.

later stove was manufactured by Musgrave & Co of Belfast, who were major suppliers of free-standing 'slow combustion' stoves for use in houses, churches and other establishments from the 1850s; their 'Ulster' stoves were notable for being lined with refractory bricks (Fig 5.6). They later moved on to manufacturing boilers and other heating equipment and continued trading until the 1960s. The flue for both of these stoves is hidden underground, which reduces slightly the visual intrusion of these appliances; a hatch in the floor provides access for cleaning. A similar arrangement exists at Calke Abbey,

Fig 5.6
A 'slow combustion stove' made by Musgrave & Co of Belfast, similar to the stove installed at The Argory. From W Eassie's Healthy Houses: A Handbook, 1872, p 186.

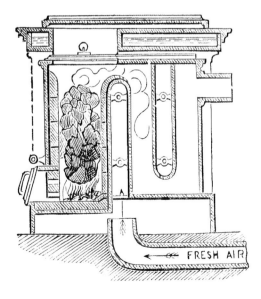

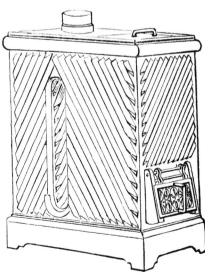

Fig 5.7
*The 19th-century cast-iron
stove in the entrance hall at
Calke Abbey.*

Derbyshire, where a later 19th-century cast-iron stove with underground flue survives in the entrance hall (Fig 5.7). This is the only heating apparatus in the house apart from open fireplaces, which is in stark contrast to the sophisticated heating systems installed in its hothouses (*see* Chapter 2). Calke is particularly noted for its lack of modernisation so it is possible that such stoves were once more common before they were replaced by less obtrusive forms of heating in houses that have been subjected to more modernisation than Calke.

The physical remains suggest that enclosed stoves were not commonly used in servants' areas of the house, either. A rare example is the small pot-bellied cast-iron stove, probably of American design, manufactured by Smith and Wellstood of Bonnybridge, Scotland, which can be found in the butler's sitting room at Attingham Park, Shropshire (although it is likely that this room is not the stove's original location).

In many houses, rooms such as butlers' and housekeepers' sitting rooms and servants' halls were equipped with small cast-iron ranges with open grates, which could not only heat the room but also boil a kettle on a hot plate or heat water in a small cistern alongside the range; examples of these 'hybrid' ranges can be seen many places, including the butler's pantries at Belton House, Lincolnshire, and Calke Abbey, and the housekeeper's room at Uppark, West Sussex (Fig 5.8) (*see also* Chapter 6). Stevenson advocates the use of such ranges in nursery suites but no surviving examples have been found.[18]

Warm-air central heating

The concept of central heating – where heat from a central source is distributed around a building – can be traced back to the Roman hypocaust, in which the flue gases from a stove

Fig 5.8
*A 'hybrid' range for
heating and cooking, in
the housekeeper's room
at Uppark.*

were passed through passages under the floors of rooms. Given the growing interest in Roman civilisation in 18th-century Britain, it might be expected that this idea would have been revived and widely used to heat country houses, but, in fact, only one example of this system has so far been identified. Dodington Park, Gloucestershire, was built for the Codrington family by James Wyatt between 1798 and 1813 and plans dated 1808 show that the flue gases from a furnace in the basement were ducted through the hollow floor of the staircase hall above.[19] James Wyatt may have adopted a similar system to heat the cloisters at Wilton.[20] The concept of the hypocaust was taken up again after the First World War by the company G N Haden & Co, with a system in which fans pumped air heated by electricity through specially designed hollow floor bricks, but no examples of its use in country houses have been found.[21]

As described in the previous chapter, textile mills provided the location for pioneering work on gas lighting and the same is also true for central heating. These mills were among the first large spaces which required heating to a reasonable and consistent level, not just for the comfort of the workforce but to facilitate the process of spinning thread. Some early textile mills were heated by stoves on their ground floors, with the flue gases carried around the upper floors in iron pipes before exiting via a chimney. By the late 18th century, many larger spinning mills were using steam engines and thus had an economical source of steam which could be piped around the building and, in some cases, injected into it, because spinning fine cotton required high humidity as well as temperature.[22] The domestic application of steam heating is described later in this chapter but the first form of factory heating which found its way into the country house used warm air.

The acknowledged pioneer of warm-air heating was William Strutt, a member of a prominent Derbyshire textile manufacturing family, who designed what became known as the cockle stove, although it is possible that this was based on ideas promoted by the Derby clockmaker John Whitehurst.[23] The cockle was a coal-fired stove located in the basement of the building, with a large, domed wrought-iron top which acted as a heat exchanger (Fig 5.9). Air passing over the outside of this dome was heated and conveyed by ducts to the upper floors of the building. Records suggest that the Strutts were employing this invention from 1792 and one

HOT AIR STOVE.

Fig 5.9
William Strutt's cockle stove, Derbyshire Infirmary. From C Sylvester's The Philosophy of Domestic Economy, *1819.*

example survives in their Belper North Mill, built in 1804.[24] William Strutt went on to employ this form of heating in his own house in Derby, and at the Derbyshire Infirmary, of which he was a patron, both around 1807.[25] Through Strutt's business connections with Boulton and Watt, cockle stoves were installed in a number of textile mills and public buildings from around 1796.

The earliest known example of the use of a cockle stove for domestic heating was at Soho House, Birmingham, an early and grand example of an industrialist's suburban villa, which Samuel Wyatt remodelled for Matthew Boulton in the late 1790s. As part of this project, Boulton installed a cockle stove in the basement which still survives, with wooden ducts hidden in the walls and under floors to convey warm air to the ground and first floors, controlled by crude butterfly valves at various points. The staircases from the ground to first and first to second floors have sets of holes drilled in every riser, from which the heated air emerged (Fig 5.10). This type of heating was subsequently promoted by Charles Sylvester, an engineer who had been employed by the Strutt family before setting up his own business. In his 1819 book, *The Philosophy of Domestic Economy*, Sylvester illustrates the system at the Derbyshire Infirmary in some detail and lists some of the other buildings heated by this system.[26] The majority of these were public buildings such as hospitals, prisons and churches but the list includes five other private residences, including Armley Hall, Leeds, home to another textile manufacturer, Benjamin Gott, and Norton Hall, Sheffield. In general, cockle stoves found more favour as a means of heating garden hothouses. Sylvester's company continued to manufacture heating equipment, eventually becoming Rosser and Russell Ltd; a Sylvester stove dating from around 1850 sur-

vives in a basement at Osborne House, on the Isle of Wight. In addition to the unusual tubular heat exchangers within its fireplaces, Wollaton Hall evidently had some form of cockle stove in its basement; warm-air ducts rise either side of the main staircase, where there is an elegant adjustable vent dated 1823, made by Harrison of Derby, the ironworks which supplied stoves and other equipment for William Strutt's house.[27] Harrison also supplied a cockle stove for one of the hothouses at Calke Abbey in 1836, as described in Chapter 2.

Another manufacturer of cockle stoves was Veitch and White, who produced a lower-cost version of Sylvester's design, which incorporated a filter to remove dust from the incoming fresh air. One of their stoves was installed at Audley End, Essex, in 1828 and they were also used in the British Museum.[28] A cockle stove heating system was installed for the Earl of Longford in Tullynally Castle, County Westmeath, Ireland, around 1807 – perhaps the first example of central heating in Ireland.[29] It was designed by the neighbouring landowner, Richard Edgeworth, who, along with his daughter, the writer Maria Edgeworth, was a friend of William Strutt.

However, from the 1820s it was the Trowbridge company G & J Haden (predecessors of G N Haden) who became the leading manufacturer of this type of stove, in which the air was heated by direct contact with the metal casing and flue of the stove, and ducted to rooms above. Brothers George and James Haden were apprentices at the Boulton and Watt Company in Soho, Birmingham, where their father was also an employee. From around 1809, George junior became one of the company's travelling engine erectors, a job which took him to the thriving textile manufacturing town of Trowbridge, where he set up his own foundry company in partnership with his brother around 1815.[30] It is highly probable that the brothers' previous employment made them familiar with William Strutt's cockle stove and the attempts by Strutt, Boulton and Charles Sylvester to transfer this technology from textile mills to domestic and public buildings; their early designs are similar to those produced by Sylvester. Entries for their own version of the cockle stove, which they called 'ventilating warm-air stoves', began to appear in their order books from 1819 and soon came to replace the mill work and other engineering components for local industries which had been the mainstay of their business.[31] They offered a small number of standard designs,

Fig 5.10
A staircase with crude heating vents at Soho House, Birmingham.

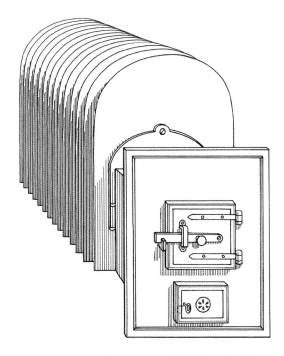

orative gratings for warm-air ducts. Later, the company supplied components for circulating hot water and steam heating systems: their name can be seen on radiators in a number of properties, such as Tyntesfield, near Bristol, and the unusual curved radiators in the stair turrets at Penrhyn Castle, Gwynedd.

An alternative type of ducted warm-air system used a coal-fired boiler located in a furnace room in the basement of the building which circulated hot water through an adjacent heat exchanger made up of metal pipes or finned plates (Fig 5.13). Such systems were manufactured by Price and Manby (also known as Price, Lea & Co) of Bristol and London from the 1830s.[35] A set of 'Price's Apparatuses' was installed in the basement of each of the four pavilions at Holkham Hall in 1844, but only the brick warm-air ducts now remain.[36] The recently-excavated basement of Worsley New Hall, near Salford, the house built for the Earl of

Fig 5.11
A warm-air stove of the type supplied by Haden and Sylvester; hot flue gases passed through the inside of the corrugated chamber, which heated the fresh air which passed around the outside. Drawing by Stephen Conlin, based on an original sketch of the stove at Dinton Park.

which were built into small brick chambers with the fire and ash doors accessible from the front. The iron casings were often corrugated to increase their surface area (Fig 5.11). Later versions also had one or more separate iron 'smoke boxes' inside the chamber – effectively secondary heat exchangers – through which the flue gases passed.

Many of their stoves can still be found in churches but the only known country house example is in a basement corridor at Erddig, Wrexham; company records show this to have been supplied in 1826 (Fig 5.12).[32] Warm air from this stove entered the rooms above via iron floor grilles in the doorways. By 1855, they had sold over 1,500 stoves, the majority going into churches and chapels, but a brochure produced around 1850 lists over 170 country and town houses in Britain and Ireland which they supplied.[33] Some large houses, as well as Windsor Castle, installed several of these stoves while others, like Erddig, appear to have had only one. During this period, in addition to their standard warm-air stoves, G & J Haden supplied country houses with drying and iron-warming stoves for laundries, ovens for kitchens and domestic hot water systems. Dinton Park (now called Philipps House), in Wiltshire, has a simple kitchen oven, a drawing of which survives in the Haden order book for 1830.[34] It also has a warm-air stove which does not bear the Haden name but is very similar to their designs (see Fig 5.11). The Haden foundry also produced a range of dec-

Fig 5.12
A warm-air stove at Erddig, supplied by G & J Haden of Trowbridge, 1826.

Fig 5.13
Price and Manby's 'patent
warming apparatus'.
From The Civil Engineer
and Architect's Journal,
July 1838.

Fig 5.13
Price and Manby's 'patent
warming apparatus'.
From The Civil Engineer
and Architect's Journal,
July 1838.

Fig 5.14
The basement at Worsley
New Hall housed the hot
water boiler, with heat
exchanger coils in the
chambers either side.

Ellesmere around 1840 and demolished after the Second World War, contains the remains of a substantial warm-air central heating system. Tubular heat exchangers were housed either side of the boiler in small brick chambers, through which fresh air ducted from outside the house was passed to heat the upper storeys (Fig 5.14). Another property with extensive remains of a warm-air heating system is Penrhyn Castle. Again, this appears to be part of the original construction of the house, which started around 1821, so this would have been a relatively early example of this technology. Drawings show that there were four separate furnace rooms under different parts of the house, three of which clearly show their original function today through the blackening of the walls. Records show that G & J Haden supplied warm-air stoves to Penrhyn Castle in 1824 and 1826 so it is possible that this type of stove once occupied some of these furnace rooms, although the rooms themselves are larger than would have been needed for Haden stoves.[37] There are circular and rectangular brass grilles in the floor at either end of the Grand Hall, each apparently supplied from a separate furnace room (Fig 5.15). An extensive wet central heating system was installed in the second half of the 19th century as a replacement for the original warm-air system.

Many country houses have grilles in the floors of principal ground floor rooms or ducts in their basements which indicate that they had some

form of warm-air heating, including Dyrham Park (Gloucestershire), Belton House, Blenheim Palace (Oxfordshire), Chirk Castle (Wrexham), Dudmaston (Shropshire), Ickworth (Suffolk) and Tyntesfield. Harlaxton Manor, in Lincolnshire, has some grilles in the skirtings and a pair of levers under the main staircase, labelled 'A' and 'B', which are believed to have controlled the flow of warm air. The burnt-out shell of Witley Court, Worcestershire, provides a splendid view down into the basement beneath the staircase hall, which housed a central furnace (supplied with coal by a railway, described in Chapter 8) with heat exchanger chambers either side. The strategically located furnace rooms and large air ducts that these warm-air systems required were hard to incorporate into the fabric of existing buildings, although this was evidently achieved at Holkham Hall. This form of heating was, therefore, predominantly employed in new houses, particularly those built or extensively rebuilt in the period roughly 1820–1870. Construction of Waddesdon Manor, Buckinghamshire, which had an extensive warm-air heating system with three furnace rooms in the basement, com-

menced in 1874, making it perhaps one of the last great houses to employ this form of heating before circulating water systems predominated.

The transition from warm-air to circulating hot water systems

Technically, it is only a small step from systems such as Price's Apparatus (also known as Price and Manby's Apparatus), in which water, heated by a boiler, circulates at low pressure through a heat exchanger in the basement of the building, to ones in which the heating pipes or radiators are located within the room to be heated, in the manner of most modern central heating systems. In fact, as we shall see, a combination of technical and philosophical constraints slowed this development. However, evidence from two properties shows that systems combining warm air and circulating hot water were utilised, at least briefly.

Wrest Park, Bedfordshire, was built 1834–9 by the second Earl de Grey, who was the first President of what became the Royal Institute of British Architects, and is believed to have designed the house himself in uncompromising French style, influenced by books on French architecture.[38] The whole property, including the service wing, stable block and kitchen garden, was built in a single phase. Wrest Park is notable for the surviving evidence of its original heating systems. Through an iron door off the basement corridor is a furnace room in two parts. One part is a boiler room, which now contains a late 19th-century boiler. The other part is a separate vaulted chamber with blackened plaster which housed the heat exchanger. In the ceiling are inlets to ducts with flaps controlled by a rotating shaft (Fig 5.16), which is connected via a chain to a control in the hall above, so the flow of warm air was adjustable from ground floor level. There

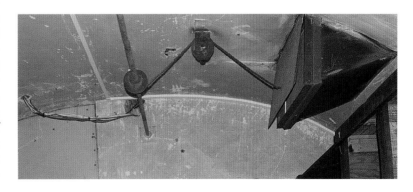

Fig 5.17
The heat exchanger
chamber beneath the
ground-floor Gallery,
Audley End. Cold air
entered via a duct in the
right-hand-side wall, was
warmed by coils of hot-
water pipes, and exited via
the hatch in the ceiling.

are a number of square and circular brass grilles in the floor of the Great Hall, from which the warm air from these ducts was emitted. These features are evidently part of the original fabric of the house but it would appear that Wrest Park also had some circulating water heating as well from the outset. There are enclosed radiators in the library and in the entrance hall and while the radiators themselves may date from later in the 19th century, the enclosures in the library are built into the bookcases, which were installed in the 1840s. A plan of the basement level shows that there was a furnace room under the northeast corner of the library, which may have housed

the boiler which heated this room.[39] Wrest Park also had a piped hot water system as part of its original design: accounts for the construction of the house show expenditure of over £1,200 for: 'heating the mansion, conservatory, hot houses, coach house, mushroom & apple house, drying sheds, laundry, stove and drying closet, two hot baths and necessary apparatus for supplying hot water upstairs, copper in the dairy scullery and press in the linen room'.[40] A more extensive circulating hot water heating system with radiators replaced the warm-air heating system some time around the end of the 19th century.

There is evidence that Audley End also employed warm-air and circulating hot water heating systems simultaneously. Under the Gallery on the ground floor is a boiler room, which was accessed via steps in the garden, and to one side is a characteristic barrel-roofed heat exchanger chamber, with fresh-air inlet ducts in the side wall and a warm-air outlet duct in the roof (Fig 5.17). It is believed that this was the location for the 'hot water apparatus for airing mansion' purchased in 1847.[41] However, the two marble-topped cast-iron enclosures in the Great Hall which housed coils of hot water heating pipes (and now house modern radiators) (Fig 5.18) are shown in a watercolour painting dated 1853 so it seems probable that these hot water heating pipes, and perhaps others on the first floor, were installed at the same time as the warm-air system and fed from the same single boiler.[42]

Fig 5.18
A radiator enclosure in the
Great Hall, Audley End,
c 1850.

The growth of circulating hot water heating systems

As the evidence from houses like Wrest Park and Audley End shows, the form of central heating with which we are most familiar today, using hot water circulating through pipes or radiators within the rooms themselves, was employed as early as the 1840s, but its adoption grew only slowly through the 19th century. This was partly a result of technical limitations: in the absence of any form of mechanical pump, these systems had to rely on natural convection to circulate the hot water. This limited the range of each system, so a large country house needed several separate systems, each with its own boiler. It also meant that large pipes, at least 50mm and sometimes 100mm in diameter, had to be used. Pumps driven by electric motors, originally know as 'accelerators', were introduced in the early 20th century – the massive under-floor and ceiling-mounted heating system installed at Eltham Palace, Greenwich, around 1935 had two large accelerators, each powered by a three-phase electric motor.[43] However, electrically driven central heating pumps did not come into widespread use until after the Second World War. Before electrically powered pumps became available, some large commercial and public buildings used devices known as 'Pulsometers', in which steam was injected into a vessel in the heating pipework to pump the water, but no examples have been found in country houses.[44]

A more powerful restraint on the adoption of circulating hot water heating was the 19th-century obsession with fresh air, described at the start of this chapter, which meant that devices which heated the air already within the room were vigorously declared to be unhealthy. Nevertheless, many houses do retain evidence of early low pressure hot water heating systems, most commonly in the form of straight lengths of pipe running in ducts within floors beneath cast-iron grilles, or as coils of pipe hidden within decorative enclosures. Unusually, Cragside has large coils of pipe in chambers below the wooden floors of some of its principal rooms with no gratings in the floors. However, almost invariably, the ducts or enclosures which housed these heating pipes were supplied with fresh air, conveyed by ducts under the floor from the outside of the building. Exceptions to this were generally confined to large, well-ventilated spaces such as entrance halls and staircase halls, until the prejudice against heating the air within rooms faded around the end of the 19th century.

To increase the area of the heated surface, fins were sometimes attached to under-floor heating pipes or short lengths of cast- or wrought-iron pipe were formed into arrays to create primitive radiators (a misnomer, since most of the heat is given off in the form of convection) (Fig 5.19). These were sometimes located unobtrusively at the bottom of main staircases, so that some heat would rise to warm the corridors of the upper storeys. From the late 1860s, purpose-designed cast- and wrought-iron radiators were introduced, initially from America, for use with both hot water and steam systems, and the following decade saw the development of the sectional radiator, in which a variable number of vertical iron sections were joined together to provide a radiator of a size to suit the space available.[45] The joints could even be made at an angle to create curved radiators, such as those in the service staircase towers at Penrhyn Castle. Some radiators were of sufficiently decorative design to be on display, especially in corridors and service areas, but more frequently they were housed in

Fig 5.19
Finned heating pipes in a duct in the floor of the Great Hall, Stokesay Court.

High pressure hot water heating systems

A widely used variant was the high pressure hot water system, patented by Angier March Perkins in 1831. His father was an American who moved to London in 1819; both father and son were responsible for a number of inventions relating to heating and other uses of steam power. The Perkins system used mostly 25mm-diameter, thick-walled wrought-iron pipe and forged fittings to form a closed system which was tested to pressures up to 200 bar. The elevated pressure enabled the water to be heated to much higher temperatures than in conventional systems: the expected maximum temperature was 177°C but the controls were crude and it is likely that this was often exceeded.[46] The compactness and efficiency of this system led to its rapid adoption in many commercial, industrial and public buildings – one notable installation was at the British Museum in 1835 – and it was evidently also tried in some larger houses.[47] A brochure of 1840 lists almost 30 country houses, and many more town houses, in which Perkins installed his system, including Dudmaston, Badminton House in Gloucestershire and Ingestre in Staffordshire.[48] However, the high operating pressures and temperatures posed a significant risk of fire and explosion. A letter to *The Times* in 1841 listed 23 properties where this had occurred, causing significant damage, including Stratfield Saye in Hampshire, home of the Duke of Wellington, and five other country houses.[49] The author of this letter is likely to have been the promoter of a rival form of heating but the hazards were real enough and many insurance companies imposed a significant uplift to their normal premiums for properties with high pressure heating systems.[50] Evidence of high pressure hot water heating systems similar to those designed by Perkins survives in a number of churches in Britain but the only example seen in a country house in the course of this research is at Stokesay Court, Shropshire, built 1889–92. The high pressure system here was replaced with a conventional low pressure system in the early 20th century, but much of the high pressure pipework can be seen in the basement, supported on rollers to accommodate expansion. A high pressure radiator also survives in the space under the main staircase, supplied with fresh air via a duct from the outside.

Fig 5.20
Decorative radiator enclosure in the dining room, Powis Castle; in the servery behind is an early electric warming cabinet.

Fig 5.21
Detail of a late-19th-century decorative radiator at Sunnycroft, Shropshire.

elaborate decorative enclosures, boxed in under window seats or built into furniture such as library bookcases (Figs 5.20 and 5.21). Excellent examples from the second half of the 19th century can be seen at the Marquis of Bute's houses: Mount Stuart, on the Isle of Bute, and Cardiff Castle, where the heating was the work of the noted engineer Wilson Weatherly Phipson; the latter property has radiators enclosed under the library tables. At Tyntesfield, a heating coil fed from the central heating boilers is fitted under the bed of the billiard table.

No boilers dating from the first half of the 19th century are known to survive, although there are numerous later examples, some of which are replacements for earlier originals. Wightwick Manor has two boilers in the basement which are probably part of the original late 1880s construction, one of which heats the Turkish bath on the ground floor. In the basement of the main wing at Osborne House are the foundations for two large horizontal boilers, with adjoining coal cellars fed from chutes in the courtyard above. It is believed that there were originally three such boilers, two for the heating system and the third supplying domestic hot water.

Steam central heating systems

Domestic central heating by steam was first proposed by a Colonel William Cook in 1745 but there is no evidence that his designs were ever employed. Both James Watt and Matthew Boulton reportedly experimented with steam heating in their homes and offices as early as the 1780s and the latter also installed this form of heating in some other houses, although it appears that these were not successful. In 1794 James Watt visited Bowood, Wiltshire, where an inventor named Green had installed steam heating for the library, but Watt was unable to cure the problems of leakage and the system was removed.[51] These systems probably followed the same pattern as the earliest examples in factories, in which the steam was simply vented to the atmosphere after circulating through the system, but later systems were closed, with the steam and condensed water returning to the boiler. One drawback of this was, therefore, the need to design the pipework so that condensed water drained back to the boiler rather than collecting in low points and impeding the flow of steam. Steam boilers needed near-constant attention to ensure that they produced steam at the required rate and the pipework carrying the high temperature steam also needed careful design to cope with the expansion and to avoid damage to furnishings etc.[52] Consequently, steam

heating was rarely adopted inside country houses, although it was more popular in public buildings, especially in America, and was quite widely used in hothouses (see Chapter 2). Sir Walter Scott, a noted early adopter of technology, including gas lighting, installed steam heating for his hothouses at Abbotsford, Roxburghshire, in the 1820s and certainly considered it for the house itself, but there is no clear evidence that he proceeded with this plan.[53] One country house which is known to have been heated by steam is Alnwick Castle, Northumberland; a late 19th-century services plan shows steam heating coils beneath the floors of many rooms, with the steam boiler house located in the gardens approximately 300m from the castle.[54]

The most extensive and sophisticated use of steam for domestic heating came later in the 19th century at Tatton Park, perhaps the most technologically advanced house in the country around the 1880s. The precise chronology of the deployment of steam at Tatton is not fully understood but it seems likely that the first pair of Lancashire steam boilers was installed in a purpose-built boiler house around 1880, to provide steam for heating around the estate and to supply a steam engine which drove a refrigeration plant (see Chapter 6) and laundry machinery; electricity generation was added in 1884.[55] There is evidence that the mansion at Tatton had initially had some warm-air heating stoves and these appear to have been replaced or

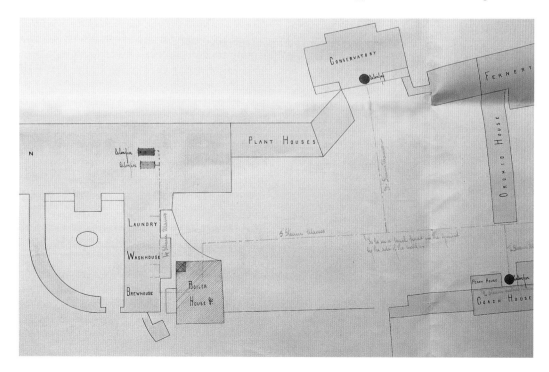

Fig 5.22
Part of the steam heating
system, Tatton Park,
showing the boiler house
and some of the extensive
steam distribution network.

supplemented with at least two separate circulating hot water systems; the extensive range of garden hothouses also had separate hot water systems.[56] In the 1880s, the two central heating boilers and at least one boiler for domestic hot water in the mansion were replaced with calorifiers heated by steam from the central boiler house (Fig 5.22). The network extended throughout the gardens, where calorifiers replaced at least five boilers in the garden hothouses and a steam pipe even ran to the estate farm, 600m north of the house, to supply a steam engine there.[57] This system must have achieved a great saving in the labour previously required to stoke all of these separate boilers and is believed to have been unique in Britain.

Use of other fuels for heating

Gas fires became increasingly popular from the last quarter of the 19th century, especially in the suburban homes of the expanding middle classes. Radiant gas fires were usually fitted into existing fireplaces, while unflued convector heaters were commonly used in entrance halls. However, even those country houses which made extensive use of gas for lighting seem to have shunned this form of heating; the only example found in the course of this research is a gas convector heater in the hall at A La Ronde, Devon, probably dating from the 1930s. It is not clear whether the failure to adopt gas heating was due to concerns about the effects of the fumes from gas fires (although most radiant fires were designed to vent their fumes via a chimney) or to some more deep-seated prejudice against this 'bourgeois' product. Certainly, electricity does not seem to have suffered the same fate: early 20th-century electric fires have survived in many country houses, including Castle Drogo in Devon, Wightwick Manor, Cragside (Fig 5.23) and Bateman's, in East Sussex. After the Second World War, most country houses installed modern oil or gas-fired central heating systems; many of these have utilised original radiators and, in some cases, such as at Penrhyn Castle, modern boilers now occupy the original furnace rooms and modern pipework often uses the original warm-air ducts to reach the upper storeys.

Ventilation

After many centuries when the objective might have been to eliminate draughts from medieval castles, halls and palaces, increasing attention was given to improving the ventilation of buildings from the late 18th century, prompted by growing scientific interest in matters such as hygiene, respiration and combustion. This was most apparent in public buildings such as theatres and assembly rooms, which were designed to accommodate greater numbers of people and were lit to a higher standard with oil and, later, gas lamps, but the same was also true to some extent for the grand rooms of country houses. Larger public buildings were sometimes constructed with dedicated vertical ventilation shafts, terminating at roof level with a cowl to increase their effectiveness, or even ventilated with steam-powered fans. No examples of these have been found in country houses, whose high chimneys were effective in drawing air from within, even when the fires were not lit, to the extent that fireplaces were often fitted with doors to restrict the air flow when not in use. Specially designed valved ventilators were sometimes cut

Fig 5.23
Early 20th-century electric fire in the study at Cragside.

into chimney breasts, just below ceiling height, to allow the chimney to extract warm air from the room.[58] The use of fires to improve ventilation was a technique which had been employed in mines from medieval times and some public buildings, including the House of Commons, had separate fires and chimneys for this purpose.[59] In country houses, however, the normal fireplace was sufficient. Gas lights – which consumed large quantities of oxygen and emitted acidic fumes, therefore contributing to the need for ventilation – could also be used to facilitate ventilation: large lights, such as the 'Sun Light' at Wimpole Hall, Cambridgeshire (*see* Chapter 4), which vented their combustion products directly to the outside of the house, helped to draw fresh air into the room.

As the words of Benjamin Franklin quoted earlier in this chapter emphasise, rooms with open fires needed ventilation to allow the chimney to work efficiently but this had to be carefully designed if uncomfortable draughts were to be avoided. Among the many devices designed to achieve this was Sheringham's ventilator, a rectangular sprung flap mounted in a duct connecting the room to the outside of the house.[60] Control flaps of this nature may be found inside the ducts which conduct fresh air from outside to grilles in the floor or walls under ground floor windows, for example at Culzean Castle, Ayrshire, and Osborne House. Another example of such devices was the Tobin tube, produced by Tobin's Patent Ventilating Company, which consisted of a vertical tube, like a periscope, fitted on the inside wall of a room, with its upper end

open to the room and lower end open to the outside of the house. The idea was that the vertical movement of the fresh air in the pipe would carry this air up to the higher parts of the room, creating circulation without excessive draught; some examples have a butterfly valve to regulate the air flow.[61] Examples of these survive in several ground floor rooms at Wightwick Manor (Fig 5.24) and Tyntesfield.

Country house kitchens and laundries were customarily built double-height, with opening windows high in the walls or ventilators in the ceiling to extract steam (Fig 5.25). Tobacco

Fig 5.24
A Tobin tube with adjustable flap in Lord Wraxall's sitting room, Tyntesfield.

Fig 5.25
The kitchen, Culzean Castle, with high opening windows to remove smoke and steam.

smoke also created a need for special measures for ventilation. Robert Kerr wrote: 'The pitiable resources to which some gentlemen are driven, even in their own homes, in order to be able to enjoy the pestiferous luxury of a cigar have given rise to the occasional introduction of an apartment specially dedicated to the use of tobacco.'[62] He recommended that these be located in a dedicated tower or ground floor annexe but went on to suggest that billiard rooms, if sufficiently separate from the rest of the house and fitted with large ventilators in the ceiling, could provide a satisfactory alternative. Weston Park has separate single-storey smoking and billiard rooms, created out of former courtyards, with ventilators in the roofs, and at Brodsworth Hall, Yorkshire, the billiard room is in a single-storey annex to the main building, again with large chimney-like ventilators on the roof.

Attention was given to the ventilation of stable blocks as early as 1625: the original stable block at Welbeck Abbey, Nottinghamshire, reportedly had adjustable ventilation shafts in each stall to carry away the horses' breath.[63] In the 18th century, stables were often fitted with ducts which conducted air from the stable area through the loft above to the roof. The ventilation cowls, which were the terminals of these

ducts and were disguised by decorative turrets, remained a feature of many new stable blocks up to the end of the 19th century (Fig 5.26). Scientific debate about the need for and methods of ventilation was at least as vigorously applied to stables as it was to houses during the Victorian era but measures to introduce fresh air were often opposed or subverted by grooms, who feared the effects on the health of their horses, or at least found it easier to produce the glossy-coated appearance of health in poorly ventilated stalls.[64] Tobin tubes can sometimes be found in stables, for example at Weston Park, and the later of the two stables at Powis Castle, Powys, has an early example of an electrically driven ventilation fan.

In conclusion

Within the overall context of country house technology, it is clear that heating and ventilation was an area which saw a relatively slow pace of change. As the above descriptions of the developments in heating systems demonstrate, from the late 18th century onwards there was no shortage of new ideas in this field, but the physical remains suggest that most country

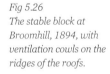

Fig 5.26
The stable block at Broomhill, 1894, with ventilation cowls on the ridges of the roofs.

houses continued to rely heavily on open fires until well into the 20th century. One explanation for this may have been the abundant supply of timber on country house estates, but the technical difficulties in balancing the needs for warmth and fresh air in large houses proved immense. For some, however, any system which provided warm air to a room rather than radiant heat was unhealthy: 'The system of warming our rooms by the air we breathe is not natural ... To make air hot enough to be the sole means of heating spoils it for breathing,' wrote J J Stevenson in 1880.[65]

Where early forms of central heating were adopted, their limited effectiveness until the advent of electrically driven pumps or fans meant that they offered little more than modest background warmth for a few principal ground floor rooms and provided no heat at all for the floors above. The proliferation of separate heating systems in larger houses added to the enormous task of keeping the house warm. At Holkham Hall, for example, one person was employed full time to stoke the four furnaces in the basement and in some other properties, such as Penrhyn Castle, it is easy to imagine the task being beyond the capability of one person. Most significantly, 19th-century heating systems did not diminish the massive servant effort required to cut and cart logs, carry coal around the house and make and clear out dozens of open fire grates, which remained the main source of heating throughout the house. It was often well into the 20th century before technology, in the shape of solid fuel, oil or gas-fired heating with electrical pumps or fans, finally created acceptable levels of comfort in most country houses and reduced, but even then did not necessarily eliminate, the servants' workload in maintaining open fires. Sadly, this dramatic transformation, eloquently described by Major Morritt, who wrote of sitting down to dinner in Chirk Castle in 1911 in his fur coat (see Chapter 1), came rather late in the life of most country houses. Felbrigg Hall in chilly North Norfolk, for example, only installed limited central heating in 1967, two years before the property was acquired by the National Trust.

6

Food preparation and storage

'It must be remembered that [the kitchen] is the great laboratory of every household, and that much of the "weal or woe", as far as regards bodily health, depends upon the nature of the preparations concocted within its walls.'

Mrs Beeton, 1861[1]

More has probably been written on country house kitchens than on any other aspect of domestic service in country houses, and this chapter will concentrate on technological developments in the storage, preparation, cooking and serving of food rather than on the ways in which servants operated below stairs. Kitchens were, of course, of great importance in the daily running of country houses and were often required to produce large quantities of food at once for certain meals, particularly as house parties became larger in the Victorian and Edwardian periods. Books on country house planning in the 19th century gave extensive advice on the layout and location of kitchens; Robert Kerr, for example, pointed out in 1864 that by the mid-19th century, kitchens had attained 'the character of a complicated laboratory, surrounded by numerous accessories specially contrived, in respect of disposition, arrangement and fittings, for the administration of the culinary art in all its professional details'.[2] The kitchen was the heart of a complex of rooms with specialised purposes, such as the still room, scullery, wet and dry larders, dairy, bakehouse, brewhouse, etc, plus private rooms for the senior members of the household (*see* Chapter 1 and Fig 1.3).[3]

Many early kitchens were sited away from the house to minimise fire risk and also to prevent cooking smells reaching the family apartments. Kerr stressed the importance of ventilation, not only to reduce the great heat given out by kitchen ranges, but also 'for the avoidance of that well-known nuisance, the generation and transmission of kitchen odours'.[4] The splendid Great Kitchen at Saltram House in Devon, for example,

replaced the original laundry and brewhouse after a fire in 1778, and so continued to be separated from the main house, and this remained in use until 1952 (Fig 6.1). Its vast hipped roof and high windows helped to provide ventilation. An even greater separation was effected at Petworth, West Sussex, where a fire, probably originating in the kitchen, destroyed the main staircase in 1714, and so the kitchens were moved to the detached service block across the yard and were generally accessed through a series of tunnels.

Where possible, kitchens were situated facing north or east to prevent sunlight adding to the great heat generated by the use of open roasting ranges. They were often built double height to aid ventilation. The kitchen at Cragside, Northumberland (Fig 6.2), for example, faces in these directions, while some kitchens, such as that at Calke Abbey, Derbyshire, were located in a basement and so did not receive any sun at all. It was also important to site kitchens with easy access to both the storage areas for supplies and the dining room. The latter was not so easily achieved as successive dining rooms were built in accordance with prevailing fashions, and long passages often separated the kitchen and dining room. At Belton House, Lincolnshire, for example, the original kitchen was in the basement but in the early 19th century the 1st Earl Brownlow had a new kitchen built at ground level outside the main house, linked to the dining room by a set of steps and a long underground passage. As described in Chapter 8, a railway was constructed along this passage to carry food; it has been estimated that this railway was probably used for at least 4,800 round trips each year.[5]

Few houses had such a facility, and servants must have spent a great deal of their time carrying serving dishes along tunnels and corridors. Dumb waiters or lifts were often installed to assist this process in the second half of the 19th century, as at Alnwick Castle, Northumberland (*see* Chapter 8). The development of heated cabi-

Fig 6.1
The Great Kitchen at Saltram, dating from the late 18th century but re-equipped in the 19th century with a cast-iron island range.

Fig 6.2
The kitchen at Cragside, with a high ceiling and windows for ventilation. The dumb waiter connects the kitchen with the scullery below.

nets in serveries adjacent to the dining room must have been welcomed by the household, inured until then to tepid food at best on their tables. These became practicable at a time when dining practices also changed. Meals prior to the early 19th century had generally been served *à la française*, the dishes to be served in each course being brought to the table all at once and served either by the head of the household or the guests themselves. From the early 19th century, what became known as service *à la russe* was introduced via France; this was the more modern system of servants being required to serve dishes in each course individually to the household and guests seated around a table. Food could therefore be brought up sequentially from the kitchens and reheated in the hot cabinets as needed.

Two major technical advances affected cooking apparatus from the late 18th century onwards. Firstly, the use of coal greatly increased. Many country estate owners possessed large acreages of timber and so wood had been the main source of fuel for both the household and the kitchen. However, many of them also began to exploit mineral reserves on their estates, some becoming what might be described as coal barons. Consequently, coal as well as wood began to be used much more within country houses. Secondly, the availability of both wrought and, more particularly, cast iron greatly increased from the mid-18th century onwards, resulting in different types of cooking range being developed. Other new technologies in kitchens included the use of refrigeration from the end of the 19th century, which made for easier entertaining. Refrigerators, combined with ice houses, allowed the annual bag of game from shooting parties and from Scottish estates to be preserved for longer. By the late 19th century kitchens often contained a variety of cooking stoves, some of them by then making use of new fuels such as gas or electricity; nevertheless, joints of meat or poultry roasted in front of an open range still formed the basis of many meals.

Food storage and refrigeration

The arrangements for storing food in country houses became more complex from the end of the 18th century, as the tastes of the household became more sophisticated. The first in what often became an extensive suite of specialised rooms, and the one usually closest to the kitchen, was the pantry or dry larder, for the storage of dry foodstuffs (Fig 6.3). This needed to be cool and dry, with good ventilation, and was often located on the north side of the kitchen wing; it also needed to be secure from vermin.[6] Vegetables were sometimes stored in the pantry but some houses had a separate vegetable larder. Next to this was often the meat larder, usually better ventilated than the pantry, with mesh windows to keep out flies, and ceiling hooks for hanging the meat; they sometimes also contained a block on which the joints were butchered, although Alnwick Castle, for example, had a separate butchery room. A few larger houses also had a separate fish larder, with stone shelves on which fish could be stored, covered in ice. The growing passion for shooting in the 19th century meant that many houses added game larders, usually detached from the main service buildings because they could be a source of odour (although those at Dunham Massey, in Greater Manchester, and Alnwick Castle were part of the main kitchen suite) and located in a shaded position (Fig 6.4). The walls of these larders largely comprised of mesh screens, to maximise ventilation and to keep out flies, and were filled with racks and hooks for hanging birds.

Fig 6.3
The pantry at
Llanerchaeron, Ceredigion.

Fig 6.4
The game larder, Lyme Park, detached from the service wing in a typically shaded location.

Buildings specifically designed for storing ice date back to ancient Mesopotamia and they were common in the Roman era. They were constructed at many medieval monasteries and they became fashionable in Britain in the second half of the 17th century, after Charles II had some built at Greenwich and St James's Palace, based on a design then popular in France.[7] The most common type consisted of a cylindrical brick-lined chamber, often dug into a shaded, north-facing slope; the chamber tapered to a drain at the bottom to prevent an accumulation of water which would accelerate the melting of the ice (Fig 6.5). They usually had two sets of doors with a short passageway between, for extra insulation. Ice was harvested from ponds on the estate to fill the chambers. Sometimes the ponds were created specifically for this purpose: at Tatton Park, Cheshire, a pond close to the rear of the house is called 'Ice Pond'.

Ice in a well-constructed, carefully filled ice house could last several years before melting completely, so many estates had several ice houses, which would be refilled in rotation. Ice collected from the ice house was usually stored in lead-lined, insulated chests, often kept in one of the larders, until it was required for preserving food or for making chilled dishes such as ice cream. Perishable food could also be stored in lead-lined, ice-filled cupboards, sometimes confusingly known as refrigerators (Fig 6.6). The ice harvested from local sources was not normally considered pure enough for direct consumption – it was simply used to chill other

Fig 6.5 (far left)
The entrance to the ice house, Tatton Park.

Fig 6.6
The ice cabinet enclosed in an oak case at Baddesley Clinton, Warwickshire.

117

Fig 6.7
A steam-powered ice machine, 1862. From Illustrated London News, *16 August 1862.*

Fig 6.8
A refrigerator at Tatton Park, dating from 1911, complete with its DC electrically powered vapour compression plant.

food items – but in the mid-19th century, an industry grew up importing ice by sea from North America and, later, Norway, which was regarded as being sufficiently pure to be used directly, to chill drinks, for example.[8] Ice houses continued in use well into the 20th century: Culzean Castle, Ayrshire, had a number of ice houses, including one built into a viaduct which forms part of the garden walls. Some of Culzean's ice houses were still in use in 1916.[9]

The principle behind vapour-compression refrigeration, the same system that is used by most refrigerators today, was established early in the 19th century but the first practical ice machines or refrigerators were not developed until the 1850s. Such machines needed a steam engine to power the compressor, so were mainly confined to commercial ice plants (Fig 6.7). However, there is clear evidence that a steam-powered refrigerator was installed at Tatton Park around 1884. This was exceptional, but the owner, Alan de Tatton, 3rd Baron Egerton, a keen engineer who had worked for a local railway company, clearly had an interest in refrigeration and served as the first president of the Cold Storage and Ice Association (later the Institute of Refrigeration) from 1899 to 1903.[10] Even when electric motors were available to power refrigerant compressors in place of steam engines, these appliances were slow to be adopted for domestic use; as late as 1912, a handbook on modern country house

design stated: 'A refrigerating machine is a thing which owners of country houses, perhaps, seldom consider as being suitable to their own special requirements.'[11] Unsurprisingly, the earliest known example of an early electric refrigerator in a domestic setting is at Tatton Park, where the large insulated chamber still retains its DC electrically powered vapour compression plant, mounted on the side; it was supplied by John Kirkcaldy Ltd of London and was installed in 1911 (Fig 6.8). Another large refrigerator chamber survives at Holkham, Norfolk, dating from around 1912, but the refrigeration plant itself has been removed.

Dairies

Spaces set aside for separating cream, churning butter and, sometimes, making cheese, existed in most large houses from medieval times, along with a dedicated staff of dairy maids, but in the heyday of the country house, from the late 18th century onwards, many dairies developed a distinctive character which was quite different from other service areas of the house. The reason for this was that the work of the dairy was one – perhaps the only – aspect of domestic work in which it was considered appropriate for the lady of the house and, sometimes, her female guests, to take an active interest, to the extent that they

Fig 6.9
The dairy at Berrington
Hall, designed by Henry
Holland and almost
unchanged since 1780.

might even indulge in some light work in the dairy themselves.[12] Marie Antoinette's dairy at Le Petit Trianon, Versailles, is often cited as the model for this but there were many earlier British examples – the earliest, perhaps, built for the Royal Family in the 1730s at Richmond Park – which probably inspired the French enthusiasm for the *ferme ornée*.[13] As a consequence of this genteel interest in the work and products of the dairy, higher standards of architectural style and decoration were usually applied to them than to other parts of the domestic offices or estate buildings. Many of the greatest architects of the age were employed to design these showpiece buildings, including John Nash at Luscombe Castle, Devon, and Blaise Castle, Bristol; Robert Adam at Knowsley Hall, Merseyside; and Henry Holland at Berrington Hall, Herefordshire (Fig 6.9), and Woburn Abbey, Bedfordshire; sadly some of these fine buildings no longer survive.[14] Robert Adam's younger brother James, along with Robert Nasmith, may have designed the striking late 18th-century dairy at Kenwood House, Hampstead.[15] Larger dairies were sometimes included in the buildings of the home farm, as at Wimpole Hall, Cambridgeshire, and Manderston, in the Scottish Borders, but other smaller examples were located in the service wings, often close to the kitchens so the produce would not have to be carried far. At Wrest Park, Bedfordshire, however, the dairy, a pavilion building with an elegant wrought-iron veranda, occupies a prominent position overlooking the gardens, far removed from the other service areas (Fig 6.10).

Fig 6.10
The dairy, Wrest Park,
which overlooks the
gardens.

Fig 6.11
The dairy, Dyrham Park,
with stone shelves and a
water fountain for cooling.

Fig 6.11
The dairy, Dyrham Park,
with stone shelves and a
water fountain for cooling.

By the second half of the 19th century, the idealised layout for a dairy, as described by Kerr at least, consisted of two rooms: one was to be designed to maintain an even temperature of around 50–55°F (10–13°C) to prevent milk going sour but still allow it to coagulate, with good ventilation and stone shelves around the walls to hold the separating dishes. The other, the dairy scullery, included a range on which milk could be heated and which provided hot water for washing the equipment.[16] In practice,

Fig 6.12
The horse gin which
drove a butter churn
in the adjacent dairy,
Myra Castle.

many smaller dairies combined these two into a single room but Lanhydrock, Cornwall, followed Kerr's guidance in this, as in other areas. The dairy has ingenious roof vents and the stone shelves around the walls were cooled with running water; the dairy scullery has a water-heated scalding range used for making clotted cream. The small, elegant dairy at Dyrham Park, Gloucestershire, has walls decorated with Delft tiles reused from an earlier building (Fig 6.11). A water fountain in the centre of the room helped to maintain an even temperature, a feature found in many 'show' dairies, including Kenwood.

Plans for the dairy at Dodington, Gloucestershire, designed by James Wyatt around 1797, show a central fountain and stone shelves around the walls, with a separate churning room and a dairy scullery. The plans show the dairy housed in an elegant classical building (later converted to the Dower House), also containing a plunge bath (*see* Fig 3.15) and a bakehouse, although it is not clear whether all these features were constructed as depicted.[17] As might be expected, mechanisation rarely intruded on the bucolic idyll that these dairies sought to create, so butter was normally churned by hand. However, a notable exception survives at Myra Castle, County Down, where a horse gin, housed in a covered structure which looms over the gardens, drove a butter churn in a dairy at the rear of the house (Fig 6.12).

Cooking on spits and open ranges

The main dish of any dinner in a country house, from the medieval period onwards, seemed to be a roast of some description. Peter Brears has argued that tender meats such as joints benefited from being cooked solely by radiant heat in a strong draught which carried away their heavy and cloying elements, while cooking joints in ovens, or metal boxes as he describes them, is a second-rate procedure since they soak in the fumes of their overheated fats.[18] This meant rotating joints of meat on a spit in front of an open fire, the fat and juices falling into a large pan below them. This method of cooking meat goes back at least to the Middle Ages and a good example of an open roasting hearth can be found in the Tudor Kitchen at Hampton Court Palace, Richmond upon Thames (Greater London), which remained in use until the 1730s and has recently been restored. The open fire here is flanked by a pair of cob irons, stands with horizontal bars projecting outwards from them, on which a series of spits running horizontally across the hearth could be supported. The spits themselves were once turned manually by scullions, a very hot task, but later small dogs were often used instead. The turnspit dog was a short-legged, long-bodied dog bred to run in a small treadwheel mounted on a wall near the fireplace which turned the spit via a pulley wheel. The open range in Alnwick Castle kitchen, even though late 19th century in date, still retains its arrangement of cob irons for horizontal spits in front of a roasting hearth (Fig 6.13). Basket spits had a cage of bars mounted on them to contain smaller joints such as chickens, as can still be seen at Burghley House, Lincolnshire (Fig 6.14). This range was still in use in the 19th century, the spits being turned by a smoke jack driving a pulley wheel, and juices from the meat were caught in a copper tray below.

Mechanical means of turning the spits had been in use at least since the 17th century. The smoke jack was a device placed in the chimney above the fire. It made use of the ascending heat to turn a vane linked to a series of gears and pulleys below to help rotate the meat on horizontal spits.[19] The pulley wheel could also be linked to one or more horizontal bars mounted across the front of the chimney breast, from which could be suspended a series of dangling hooks for smaller joints of meat hanging vertically rather than turning horizontally in front of the open fire.

Fig 6.13
Cob irons at the roasting hearth in the kitchen redesigned by Salvin, Alnwick Castle.

Fig 6.14
The basket spit in the kitchen, Burghley House.

Fig 6.15
The roasting hearth with horizontal and vertical spits driven by a smoke jack, at Petworth House. Parts date from the early 19th century and it was updated by the firm of Clement Jeakes in 1872.

Fig 6.16
A spit driven by a small water turbine in the room below, at Cragside.

An open range of this type can still be seen at Petworth House, dating originally from the early 19th century but updated in 1872 when other parts of the kitchen were modernised by the firm of Clement Jeakes. The smoke jack powered four dangling spits and up to eight horizontal ones (Fig 6.15).

Other mechanical devices for spit-turning were occasionally used. Chatsworth, in Derbyshire, seems to have made use of a small waterwheel to turn the spit, powered by a conduit from a pond on the hill above the house, still known as 'the jack pond'.[20] Inevitably, at Lord Armstrong's Cragside, the spit in front of the 19th-century open range is driven from a small water turbine, or Barker or Scotch mill, placed in the room below the kitchen (Fig 6.16), and there may well have been another of these at Alnwick Castle. Vertical spits driven by jacks incorporating weights and a series of pulleys were also used for smaller establishments; after the weight had been wound up, it dropped slowly and so turned the spits. The bottle jack was a variant of this, consisting of a clockwork motor inside a brass container shaped like a bottle. From this was suspended a flywheel with a vertical spit below, which enabled the attached joint to rotate; it was even possible to attach small pieces of fat to hooks on the flywheel to baste the roasting joint (*see* Figs 6.17 and 6.18). Clockwork or spring-driven jacks like this could also be enclosed in a hooded metal screen which, when placed in front of an open fire, speeded up roasting as heat was reflected from the screen back onto the joint. Mrs Beeton, however, complained that the flavour was then more like baked than roasted meat.[21]

Food other than spit-roasted meat, such as stews and sauces, was also required, and some cooking was done, at least from the 16th century, on stoves heated by charcoal. Again, an early example can be seen at Hampton Court – it is brick-built, with fire bars set into circular apertures in the top of the stove which are heated from burning charcoal beneath.[22] Three of the five arches contained ash pits linked by flues to the fires, while the other two served as charcoal stores. The fumes from burning charcoal were unpleasant and so these stoves were often placed directly below a window or provided with smoke hoods. Elizabeth Raffald, in a later edition of her *The Experienced English Housekeeper*, originally published in 1769, proposed that coals and embers should be used rather than charcoal, which she said was 'pernicious to cooks', as well as being expensive. She also recommended the use of ducts from the grates leading to flues joining a chimney, which would help to create the necessary draught.[23]

Cooking was often done on trivets raised a few inches above the tops of the grates, which could also act as grills for small meat items with the addition of a gridiron. Generally, soups, stews and sauces were made on the charcoal stoves, which could also, of course, be used for heating water and might well have a swing-arm crane positioned alongside to enable heavy pots to be lifted off more safely. Some of these stoves survive, often converted to other forms of fuel or used for other purposes, such as those at Shugborough, in Staffordshire, Dudmaston, in Shropshire, and Burghley House.

The development of the closed kitchen range

By the second half of the 18th century, the increased use of both cast- and wrought-iron, together with coal as a fuel, led to some far-reaching changes in kitchen technology. Coal would not burn successfully on an open hearth, such as had been used for spit-roasting, and so new enclosed systems had to be devised which created an updraught through a chimney to enable a coal fire to draw and so become hot enough for cooking. This led to the development of kitchen ranges as we know them today.

The open range was still intended mainly for roasting but the fire was contained in a grate of wrought-iron bars, secured by tie bars to the back of the fireplace. Often trivets or circular iron rings big enough to support a small pan or kettle were included so that they could be swung over the fire when needed. Some fires were enclosed with cast-iron panels each side, and the next logical step was to replace these with iron boxes to serve as ovens. The earliest patent for these was taken out by Thomas Robinson, a London iron founder, in 1780, and soon afterwards another iron founder, Joseph Langmead, took out a patent for a boiler alongside the grate.[24] The open range then consisted of an open fire in a grate with an oven on one side and a boiler for hot water on the other, the latter often equipped with a tap for drawing off the hot water (Fig 6.17). Since the fire was smaller than it had been previously, the meat often had to be roasted vertically and a bottle jack in a hooded screen was often used, or a vertical dangling spit.

The next major step in the transformation of kitchen ranges was the development of the closed range in the early 19th century, which was much more versatile in use than the conventional open range. In 1796, Benjamin Thompson, Count Rumford, an American-born scientist (*see* Chapter 5), criticised the open range for the amount of heat that escaped up the chimney. He proposed the use of an enclosed stove with numerous fireplaces, all of which directed heat via flues to the undersides of different cooking vessels so that none was wasted. Although impracticable in actual use because it would have had to be enormous to incorporate the number of flues he thought desirable, his principles were shortly afterwards incorporated into a patent for what is probably the first closed range, made by George Bodley of Exeter in 1802. In his stove, the fire grate was covered with an iron plate which stopped hot draughts going up the chimney until after they had been utilised in flues surrounding a hot water boiler or oven. The iron cover itself served as a hot plate for stewing meat and so on.

The initiative then passed to the iron-making district of the West Midlands, where an iron founder based in Leamington Spa, William Flavel, began manufacturing closed stoves in the

Fig 6.17
An open range with hobs each side at Tredegar House, with a bottle jack in front.

1820s. These became known as 'patent kitcheners' because of their versatility, or often as 'Leamingtons' because of the location in which they were first manufactured. The fire grate was much smaller than that of an open range and had an ash grate underneath, while dampers were used to direct the heat of the fire where needed, either to hot plates on top or the oven and hot water boiler on either side. Circular plates could be removed from the top plate for faster boiling. Later models incorporated a roasting oven and a pastry oven, in which the flues directed heat under the base to produce the bottom heat needed for baking.

Closed ranges, as well as being more economical in their fuel consumption than open ranges, were also much cleaner to operate, and pots and pans were not blackened by contact with an open fire. J C Loudon referred to a closed range by Browns of Luton and argued, 'There can be no doubt of the improvement effected in this range of the avoidance of smoke and dust, economy of fuel etc over the common range.'[25] These stoves were not so suitable for roasting, however, and since it was still thought that open-fire roasting produced a better result than oven roasting, some manufacturers made the grate bottom adjustable so that it could be lowered to increase the size of the fire for roasting. Flavel built the Eagle Foundry in Leamington Spa in 1833, which enabled the firm to increase its production, and the Kitchener was exhibited by William's son Sidney at the 1851 Great Exhibition in Crystal Palace, earning a Gold Medal, and the design was copied by many iron founders all over the country.

Small closed ranges were also often installed in rooms around the main kitchen, such as the housekeeper's room or the still room, as at Petworth or Dunham Massey, where a Trident

Fig 6.18
The Eagle cast-iron range in the old basement kitchen at Chastleton. It was installed for John H Whitmore-Jones after he inherited the house in 1828.

range was fitted in 1906 by the firm of Clement Jeakes.[26] Such rooms, originally used for distilling herbs etc, were used for making jams, jellies and other smaller culinary items by the Victorian period (*see* Fig 5.8). Early closed ranges were set in brickwork and the efficiency of the flues depended on the skill of the bricklayer. The Eagle Foundry produced wholly cast-iron ranges in which the flues were contained within the body of the ironwork. The contrast can be appreciated by comparing the range in the old basement kitchen at Chastleton, Oxfordshire, installed in the early 1830s, with the much later Gold Medal Eagle Range at Uppark, West Sussex, dating from about 1895 (Figs 6.18 and 6.19). Incidentally, Flavel's iron foundry in Leamington Spa continues to produce the Rangemaster cooker, successor to the Kitchener, while the Coalbrookdale Foundry in Shropshire, which also produced cast-iron ranges, is still the centre for the production of Rayburn cookers, a design first manufactured in Scotland in the 1940s, and Agas, originally of Swedish origin in the 1920s. Examples of these types of 20th-century ranges can still be seen in some country houses, where they replaced

Fig 6.19
The Gold Medal Eagle range in the kitchen at Uppark, West Sussex. The range was installed c 1895.

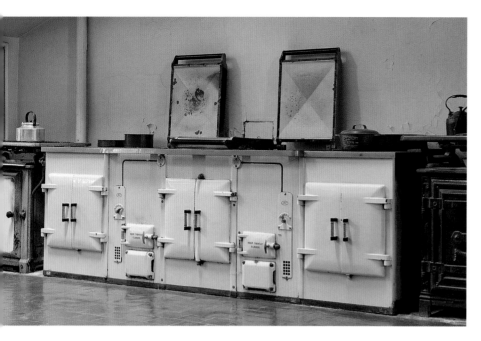

Fig 6.20
The Aga in the kitchen at Dunham Massey, with gas stoves either side.

these were rare in England although common in Germany. He said rather disparagingly that:

> The most fundamental difference between the English and the continental ranges is that the former is almost always situated in an alcove. We definitely prefer a kitchen-stove to be free-standing, at least on two sides; indeed, we think that the best form is entirely free-standing, accessible from all sides. The English, however, with unshakeable obduracy, cling to their practice of wedging it between walls.[27]

He suggested that this was due to the English passion for the open fire and that consequently they wanted to situate their kitchen ranges where the original kitchen fireplaces had been. He may well have been right as even the later ranges tended to be situated against a wall, where it was easier to install the necessary flues. English cooks seem to have preferred a central working surface for a variety of tasks rather than a central stove (Fig 6.21). Nevertheless, there is a splendid example of an island range at Saltram House in Devon, made by Flavels of Leamington Spa in the 1880s, from which the smoke and fumes were carried off in flues under the floor (*see* Fig 6.1).

earlier cast-iron ranges, such as Dunham Massey, where a huge Aga was installed in 1936 to replace an earlier closed range (Fig 6.20).

In many large kitchens of country houses, central ranges similar to the popular island cookers of today were sometimes added in the late 19th century. Stephen Muthesius, the German author of *The English House* (1904), pointed out that

Fig 6.21
The pastry chef and his staff in 1910 at Waddesdon Manor, Buckinghamshire.

Cooking by steam, gas and electricity

Another innovation was the use of steam for cooking, which had been introduced at least by 1825; the notorious Yorkshire mill owner Anne Lister wrote of a visit to Chatsworth in that year: 'The kitchen & offices quite as well worth seeing as any part of the building. Surrounded with stoves *à la française*. At one end a hot table for dishing up upon. A hot closet & steam kitchen all heated by steam.'[28] The pressurised steam was generated in boilers and conducted to a variety of kitchen equipment. In the Great Kitchen at Tredegar House, Newport, there is a late 19th-century copper steam boiler with a Bourdon pressure gauge which seems to have been used for generating steam to clean greasy pots and pans (Fig 6.22). Probably the finest late Victorian kitchen making extensive use of steam is that of Petworth, refitted after a fire in 1871, with much of the equipment supplied by Clement Jeakes of Great Russell Street, London. The coal-fired steam generator was placed in the adjacent scullery, and generated steam for a bain-marie for gentle cooking for sauces etc, and for a series of steam kettles for cooking vegetables and puddings, as well as a warming cupboard (Figs 6.23 and 6.24).

Fig 6.22
The late 19th-century copper boiler in the Great Kitchen at Tredegar House.

Fig 6.23 (below left)
The coal-fired steam generator in the scullery at Petworth House.

Fig 6.24
A steam bain-marie at Petworth, installed by the firm of Clement Jeakes.

Although gas was originally introduced into country houses for lighting (*see* Chapter 4), it was occasionally used in the kitchens, although the innate conservatism of landowners, and their cooks, and their attachments to meat roasted

before an open range restricted its wider adoption. London's Reform Club was one of the first places to install gas cookers, in 1841, and several gas cookers were displayed at the Great Exhibition of 1851. The cause of gas cooking was no doubt enhanced when a gas-fuelled pastry oven was installed in the kitchens at Windsor Castle around 1870; the remains of this survive in the National Gas Museum's collection in Leicester. The convenience of gas hobs gained them acceptance, particularly in kitchens of the middle classes, but gas ovens did not become widely popular until after the invention of the thermostatic control in 1923. At Burghley House, the hob of the cast-iron solid fuel range was converted by the firm of Benhams of London to use gas, and a large gas hob was installed by the same firm at Petworth, probably as part of the 1871 rebuilding (Figs 6.25 and 6.26). Penrhyn Castle, Gwynedd, has an unusual gas hob, probably dating from the early 20th century, with around 20 burners, some of which were for grilling.

Fig 6.25
A cast-iron stewing range converted to gas, Burghley House.

Fig 6.26
A gas range installed by Benhams of London in the kitchen at Petworth.

Domestic gas cookers can sometimes be seen alongside other cooking devices, as at Dunham Massey, where two gas cookers flank the later Aga, possibly for use when the range was not lit (*see* Fig 6.20). Electric stoves followed slowly, and a huge electric oven installed by Benhams in the 1920s can also be seen at Petworth, although apparently its electricity consumption was so great that it was rarely used. The kitchens here show clearly that it was not unusual for the cooks in country houses to be working on, in the same kitchen, 18th-century open ranges, Victorian closed ranges and even gas and electric cookers and steam-heated bain-maries. Even at Cragside, the Eagle closed range by H Walker and Sons of Newcastle and an open range by Dinning and Cooke of the same city were installed side by side in the kitchen so that cooks could roast in front of an open range, using the spit driven from a turbine, but also make use of the greater flexibility of the closed range (*see* Fig 6.2).

Baking bread had generally been carried out in standard brick-built bake ovens, dome-shaped structures rather like modern pizza ovens. These were heated by faggots of wood placed in them, and sometime below them as well, and withdrawn before the bread dough was placed in the oven. They also needed reheating in between each batch of loaves, and were usually situated in separate bakehouses as they were rather dirty to operate. This remained the practice even when more modern bake-ovens were installed, such as the one by Feethams at Gawthorpe Hall, Lancashire (Fig 6.27).

Finally, once the food had been cooked, efforts were made, at least by the 19th century, to keep it warm for service. Wooden hot cupboards lined with zinc or tin were a common sight in kitchens. These were placed in front of the open range, where they served a dual purpose of warming food containers and plates and helping to keep the vast heat from the range away from the kitchen staff (Fig 6.28). Once means of heating water became common in kitchens, particularly after the introduction of closed ranges made it impossible to heat plates in front

6.27
A bread oven by Feethams at Gawthorpe Hall.

Fig 6.28
A moveable plate and food warmer, Dyrham Park; to the right of the range is a water-heated hot cupboard.

Fig 6.29
A steam-heated warming
cupboard in the servery
at Lanhydrock.

Fig 6.30 (far right)
A charcoal-heated hot
cabinet in the dining room,
Weston Park; a matching
lead-lined cabinet used ice
to chill wine.

of the open range, separate hot cupboards heated by hot water, steam, gas or, later, electricity, became available. Some hot cupboards were placed not in the kitchen but, sensibly, in the serving area next to the dining room, as at Lanhydrock, or even in the dining room itself, where they were often concealed in elaborate cabinets (Fig 6.29; *see also* Fig 5.20). At Weston Park, Staffordshire, one of the two ornamental cabinets was heated by charcoal as a warming cupboard and the other served as a wine cooler (Fig 6.30). These developments helped to ensure that – at last – the household and guests could have hot food served to them.

In conclusion

As seen in Chapter 1, the entertainment of friends and, in some cases, important clients became one of the main functions of the country house in the second half of the 19th century. The kitchen gardens, as seen in Chapter 2, were producing greater varieties of fresh produce, while the popular practice of shooting game birds and animals, either on the owner's estate or on those of his friends, resulted in large quantities of meat which had to be stored and prepared. Technological developments in the form of much more readily available coal and cast and wrought iron came to the rescue in enabling more efficient cooking apparatus to be developed, but better methods of chilling food had to wait until the end of the 19th century.

What is perhaps surprising about all these developments is that the old was installed alongside the new; the senior kitchen staff in particular seem to have been reluctant to abandon well-tried and tested methods, such as spit roasting in front of an open range, for wholly new systems but were prepared to try these out alongside their traditional equipment. The result was the complex 19th-century kitchens which are often one of the main sources of fascination among today's visitors to country houses. But, as in the case of laundries as referred to in Chapter 3, care has to be taken to distinguish original fittings from those imported for display purposes.

7

Communications: bells and telephones

'Miseries Domestic No. 34: Pulling at an elastic bell-rope, which you either break from the cramp, without sounding the bell, or tug repeatedly (thinking that it does not ring, when it does), so as to bring up a wrong servant.'

James Beresford, 1806[1]

Chapter 1 of this book described how large service wings were added to many country houses from the late 18th century onwards, while country house planners in the 19th century stressed the idea of 'multifariousness', the provision of small rooms for particular domestic purposes such as shoe-cleaning or clothes-brushing (*see* Fig 1.3). In addition, planners devised ways of keeping the servants out of sight in the course of their duties as far as possible, burying them in basement walkways or service tunnels (*see* Figs 1.7 and 8.1). Such developments must have meant it was far more difficult to find servants to undertake specific tasks when required, a far cry from the medieval manor house where the existence of the communal hall meant that servants were generally within sight or at least easily contactable. The vast extension of the service areas of country houses could hardly have succeeded unless technology had not developed alongside these changes, firstly with the introduction of sprung service bells, followed much later by speaking tubes, electric bells and finally telephones. These were all originally intended for internal communication; the possibility of communicating directly with the world outside the country estate by technical means was only achieved towards the very end of the 19th century.

Sprung bells

In his plea for the preservation of more Victorian country houses put forward in the *Victorian Country House*, first published in 1971, Mark Girouard refers to 'rows of bells rusting in the back corridors'.[2] Fortunately, the owners of many country houses have begun to realise the importance of these fittings, and they have been made known to the general public by their prominence in the television series *Downton Abbey*. Several National Trust properties, such as Dyrham Park in Gloucestershire, Attingham Park in Shropshire and Tredegar House, Newport, have put a lot of effort into the conservation of their sprung bells, and many more are realising their importance in helping to understand the working of the household in the past. The use of service bells operated by wires dates back to at least the middle of the 18th century, a time when separate servants' quarters were being built. Donal McGarry has drawn attention to the early depiction of two bell pulls each side of a fireplace in Johann Zoffany's portrait of Sir Lawrence Dundas and his grandson, at his London home in 19 Arlington Street, painted in 1769.[3] The Braybrooke Papers for Audley End, Essex, provide evidence of the installation of bells from the 1760s onwards, with more being added from the 1780s.[4] By the time that J C Loudon referred to the art of bell-hanging in his *Encyclopaedia of Cottage, Farm and Villa Architecture*, first published in 1833, bells must have been commonplace in a variety of domestic dwellings.[5]

The bells themselves were usually mounted at cornice level, each bell hanging from a coiled spring so that a pendulum also suspended from the spring would continue to move even after the bell had stopped ringing. This gave servants time, on hearing a bell ringing, to move from more distant rooms in the service areas in order to see in which of the family rooms they were required. The use of these coiled springs seems to have been quite early in date: the household accounts for the Braybrooke family at Audley End for 1786 include entries for 'new best scrolled spring carriage to your bell fixed in passage north wing' and 'new bell with best scrolled spring and carriage and fixing in passage at top of great stairs south wing'.[6] Loudon's description of bell-hanging points out the essential elements of a sprung bell system:

The art of bell-hanging may be described as the art of conducting lines of wire, intended to ring a bell at one end, when pulled with a little force at the other, in all directions round the apartments and through the walls of a building, in such a manner as not to obtrude on the view. This is effected with ease in straight lines; and angles are got over by what are called cranks, of which there is a variety of sorts for external and internal angles.[7]

The bell wires needed to be maintained in tension to ring the bells, and each wire, as Loudon indicates, was attached to a corner of a triangular crank, made of brass or zinc alloy, a simple yet clever device that enabled the wires to be turned through 90 degrees (Figs 7.1 and 7.2). The Braybrooke accounts also contain many references to copper wire, cranks and chuck springs.[8] Copper wire, less affected by damp than iron wire, was preferred for bell wire and was generally in use by the 19th century, the gauge usually adopted being No 14 for outdoor and No 16 for indoor work.[9] It is likely that in the late 18th century, bell pulls in the family

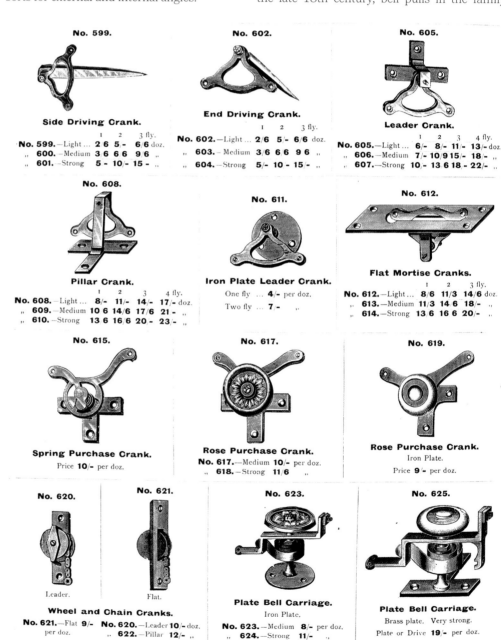

Fig 7.1
Bell cranks from the catalogue of R Melhuish, Sons & Co, London 1903.

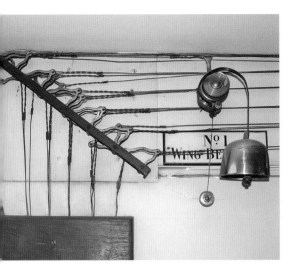

rooms were connected directly to specific bells placed in strategic locations rather than to one or more bell boards as was the case later. The bells fitted in 1785–6 at Audley End were positioned in places such as the top of the main stairs, outside the male servants' room or the north passage in the servants' quarters. In other cases, bells were positioned to summon particular servants, as in the wiring of '1 very long pull from my Lord's dressing room to bell in valet's room'.[10] By the mid-19th century at least, it seems to have become more common to group the bells together on one wall or attach them to a bell board located in the service area rather than scattered throughout the house; most of the bells at Audley End seem to have been regrouped on a bell board by the 1820s.[11] The wires themselves then needed to travel in either horizontal or vertical directions to keep the tension, and often wires were taken from the bell pulls in each room up into the attics, across the attic floor and then bunched together in metal tubes down to the bell boards. Bell wires and cranks can therefore often be found in unexpected places throughout the house. If bell systems were added to existing properties, as was very often the case, much damage could be done to existing plasterwork and so wires and cranks can still often be found at cornice level where bell pulls were used, or concealed behind dado rails where these existed. This can be seen, for example, at Traquair House in the Scottish Borders, a tower house dating from as early as the 12th century, into which bells were installed, possibly on the advice of Sir Walter Scott of nearby Abbotsford, Roxburghshire, in the early 19th century (Fig 7.3). Elsewhere, in new houses, bell wires could be grouped together in pipes in the walls and then

plastered over, as can be seen at Woodchester, Gloucestershire, where copper pipes to carry bell wires were attached directly onto the stone walls of several rooms, with the stone vaulting ribs fitted over the top of them.

Each bell was usually labelled with a painted room name on the board to which it was attached. These could be repainted as the uses of rooms, or the occupants of rooms, changed, and so the current names do not necessarily indicate when the bells were first introduced (Fig 7.4). Some sets of bells only bear numbers rather than room names, while at Attingham Park, for example, the rooms were named on the white ceramic disc forming the pendulum of each bell or, in some cases, just labelled with letters of the alphabet. It is therefore very difficult to identify the date of the installation of sprung bells just from the room names now associated with them.

Fig 7.2 (left)
The bell cranks for the row of bells in the attic corridor, Manderston House.

Fig 7.3 (below)
The bells and bell cranks in the front hall of Traquair House, Scotland.

Fig 7.4 (bottom)
The line of sprung bells, complete with discs and room labels, in the corridor next to the still room at Dunham Massey.

Many large houses had more than one set of mechanical bells, carefully placed to speed up the response time by the servants. The main set was usually placed in the main service corridor, often close to the servants' hall or the butler's pantry so that one of the senior servants could ensure that the bells were being answered within a reasonable time. A spectacular example of this can be found at Manderston House, in the Scottish Borders, rebuilt by Sir James Miller in the first decade of the 20th century, which boasts 56 bells in the lower service corridor in two sets of 28 each, one set probably dating from the enlargement of the house in 1903–5 (Fig 7.5). Another set of bells was often placed in attic corridors close to the servants' bedrooms so that housemaids and footmen could be summoned quickly for the more domestic rooms, as is the case at Manderston, where another set of 26 bells in an attic corridor are labelled exclusively for bathrooms and bedrooms, as well as WCs. At Calke Abbey, Derbyshire, there are sets of bells on the ground floor and at the top of the stairs into the kitchen, and also another set in the first floor servants' corridor, mostly labelled for the bedrooms and nurseries. Similarly, at Ickworth, Suffolk, an extensive mechanical bell system is split in two, with one set of bells outside the servants' hall in the rotunda basement and another set near the butler's and housekeeper's suites in the East Wing. The locations of bells like these are useful in trying to reconstruct the daily working of the household.

The bells themselves may be of different sizes: those for the various doors of the house were often larger than the rest of the bells and so would emit a different note, meaning that the butler or a footman could get there quickly. Tyntesfield, near Bristol, for example, has two much larger bells labelled front and back yard, while Audley End has a very elaborate front door bell (Fig 7.6). Obviously, if servants could learn to recognise the notes emitted by different bells, it would save them the journey to the bell board and enable them to go straight to where they were required. Stevenson suggested in 1880, 'Some pains should be taken to make bells different in tone, so that each is recognisable – from the deepest, which is usually the hall bell, to the highest in tone ringing from the attics.'[12] Sprung bells were still being installed in the early 20th century, when the author of an article on bells in a technical journal also argued, 'The other bells may also be selected of varying tones, to that they can in time be distinguished from each other.'[13] It is unlikely, however, that this was always the case: physical and aural tests carried out by the authors on the bells at Attingham Park indicated that only perhaps four bells of the 44 were sufficiently distinct to be readily identifiable by sound alone, including the front door bell. The tests also revealed that the bells themselves stopped moving typically within 40 seconds of the bell being rung, but the pendulums attached to the bells continued to swing for at least five minutes.

Houses exhibit a great variety of bell pulls and levers by which the bells were operated. It would appear that the earliest bells were fixed at cornice level and rung by means of bell pulls, embroidered strips of cloth or narrow pieces of tapestry with a ring pull or tassel on the lower end. In bedrooms, such pulls were usually hung each side of the bed, so enabling the occupants to summon a servant without leaving their bed. Even if the bell pulls do not remain, traces of the cranks often still survive, enabling earlier positions of beds to be worked out. Some houses continued the use of bell pulls even in the state rooms, as, for example, at Audley End, where the single bell pull in the Saloon allegedly made use of the tassels from Sir John Griffin Griffin's uniform (Fig 7.7), shown in Rebecca Biagio's portrait in the same room. The crank is neatly hidden in the decorative capitals of the pilasters, as are those in other state rooms such as the library. At Traquair House, the embroidered bell pulls were retained as being in keeping with the original period of the house, even though the bells themselves are mid-19th century in date (Fig 7.8). In other types of room, particularly those with a dado, a bell handle or lever could be fixed into the dado rail, or directly into the wall, at a level accessible by someone seated in a chair. Behind each bell handle was a chain drum in a wood, iron or brass box concealed in the wall on which they were mounted and so hidden from view: the chain was connected to the lever enabling it to operate. In many cases, these bell handles were supplied in pairs so that when fixed each side of a fireplace, the lever could be moved either to the right or the left. The wires from these usually went to a particular set of bells: many were labelled 'Up' or 'Down', indicating that one set of wires would be taken directly to the attic level to summon a housemaid, for example, or to the bell board in the basement from the lever labelled 'Down' so that a footman could be summoned to make up the fire. Even more specifically, the library at Wrest Park, Bedfordshire, for example, has two bell levers labelled respectively 'Servants Hall' and 'Groom of the Chambers'.

The bell handles themselves could be ordered to suit a room's decorative scheme, and the

Fig 7.7 (far left)
The bell pull in the Saloon at Audley End, probably dating from the 1780s and reputed to have made use of the ornamental tassel from Sir John Griffin Griffin's dress uniform.

Fig 7.8 (left)
The tapestry bell pull in the housekeeper's sitting room at Traquair House.

exposed portions were often made of porcelain, china or cut glass, and heavily gilded; others were of bronze or brass (Figs 7.9 and 7.10). In 1785 at Audley End, Messrs Willerton and Green, Jewellers, New Bond Street, were paid £1 4s 5d for an ivory and gold bell handle.[14] A particularly attractive bell handle in the Portico Bedroom at Manderston incorporates a Wedgewood intaglio, while many other bell handles were specifically made to incorporate the family crest, as at Tatton Park, Cheshire (Fig 7.11). Some particularly unusual bell levers survive at Belton House, Lincolnshire, in the front hall, designed in the shape of lions so that the lever

mechanism was worked by depressing the front paws (Fig 7.12). At Brodsworth Hall, Yorkshire, the building accounts include the following detail for the sprung bells, which were installed when the house was built in the 1860s:

Provide and fix 36 Bells and Indicators with 36 common pulls from bedrooms to passages of Servants' Offices and 8 china lever pulls from Nursery and Bathrooms.

2 Bells and pulls to Servants' Bedrooms. 1 bell and lever pull from Butler's Pantry and 1 from Housekeeper's Rooms. 2 Bells and 2 china levers from School Room. 7 bells and

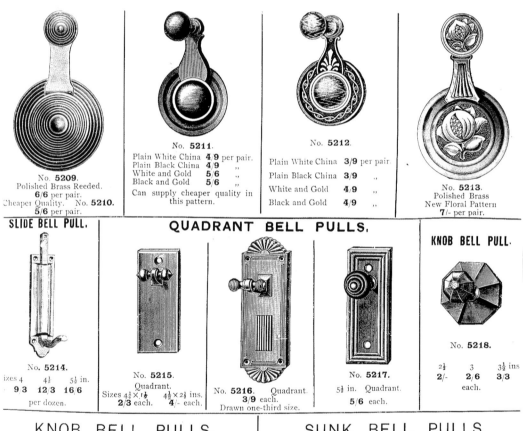

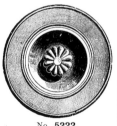

Fig 7.9
Bell levers made by Pryke & Palmer, London, c 1900.

14 China ornamental pulls of the p.c. value of 12/6 each – 2 Bells with 1 brass Call pull each from the bottom to the top of the house. 1 large alarm bell of the value of Ten pounds.

Take down and refix the Dinner Bell.

A large Bell and 2 Bronze Pulls to entrance door. 2 iron pulls and 2 Bells to Kitchen Entrances.[15]

These extracts imply that there were at least three bells for each of the 13 bedrooms or dressing rooms, presumably two of them positioned each side of the beds, with the more ornamental

china pulls in the downstairs rooms as well as some bell levers. Little survives of this bell system at Brodsworth Hall as the house was later fitted with electric bells; the board for the 36 bells can still be seen in the upper corridor of the servants' quarters, where the female staff slept, but nothing is left of the bells.

Bell-makers and fitters

It is not easy to discover who manufactured and fitted the bell systems. Given the wiring required, some of the early bell-fitting was carried out by whitesmiths, who traditionally worked with white metal such as tin or galvanised iron. At Audley End, John Pratt, referred to as a whitesmith, both provided fittings and hung the bells and wires in the 1780s.[16] However, the specialist trade of bell-hanging appeared at about the same time, with advertisements appearing in local newspapers and documents referring to bell-hangers. It is even more difficult to find out where the bells were cast. The large number of bells at Manderston in southern Scotland were cast by John Bryden and Sons, who also cast church bells, although many of these may well have been cast in the Gorbals Foundry in Glasgow.[17] Presumably other domestic bells were also cast by traditional bell-founders. How far the house owners were responsible for specifying the locations of their bells is equally difficult to find out, but Robert Stayner Holford, who commissioned the architect Lewis Vulliamy to remodel both Dorchester House, Westminster, and Westonbirt, Gloucestershire, wrote a long and detailed letter setting out his requirements for his bell system.[18] Others must surely have done the same, although by the later 19th century, large-scale contractors such as Clement Jeakes and Co were including bell-fitting among the range of domestic services they provided.

How quickly the bells were answered is often a matter of complaint in the literature of the period. *The House Servants' Directory* of 1827 points out that the answering of bells is part of a house servant's business that requires a great deal of attention:

> Whenever your parlour or drawing room bell rings, lose no time in going to answer it; never wait to finish what you are about, and leave the bell unanswered; you should never let a bell ring twice if you can possibly avoid it, for it seems to be a great part of negligence in a

Fig 7.10 (left, top)
An ornamental bell lever toning in with the decor of the Pompeii Room, Ickworth.

Fig 7.11 (left, middle)
A bell lever ornamented with the crest of the Egertons, Tatton Park.

Fig 7.12 (left, bottom)
One of a pair of ormolu bell levers, modelled as lions, in the Red Drawing Room at Belton House; the bell was rung by pressing down on the lion's paws.

Chapter 4), Lord Salisbury at Hatfield House, Hertfordshire, was to the fore in installing some of the first electric bells. William Butterfield, a carpenter and joiner at Hatfield House, who compiled a small book about alterations to the house since 1868, said that before 1869:

> The ordinary bell pulls were in use throughout the house. Very few English people, apart from the Railway Companies, had taken up electrical work of any kind, consequently electric bells were somewhat of a novelty in England. Two Frenchmen were engaged from a firm in Paris who with the aid of two or three of the Hatfield House workmen, fitted up the whole house with electric bells. The servants' rooms, which had before this been known as attics, were named to correspond with the various visitors' rooms, and fitted with electric bells to ring from the corresponding rooms: a decided improvement on the old method.[22]

Chatsworth House, Derbyshire, was not all that far behind, as the electrical bell system was there by at least 1880 from documentary evidence.[23] Frederick Allsop wrote articles on bell-fitting for the *English Mechanic* in 1887 and republished these as *Practical Electric Bell Fitting* in 1889, referring to the 'great and increasing demand for electric bells in this country'.[24] The early use of these bells was made possible by the invention of the battery, of which the most widely used was the Leclanché cell, patented in 1866. The early form was usually a glass jar full of a solution of ammonium chloride in which were suspended a zinc rod, which was positive, and a porous pot containing a plate of carbon surrounded by a mixture of crushed carbon and peroxide of manganese, the negative element. The liquid solution acted as the electrolyte, permeating through the porous pot to make contact with the carbon rod, while a copper wire outside the jar connected the two plates and so set up an electrical circuit. Later forms dispensed with the porous pot, the carbon and manganese being moulded into solid blocks and placed each side of the carbon plate. Each cell provided a voltage of 1.4 volts, and so cells were often connected in series to obtain the necessary voltage. Several country houses still have battery boxes containing the Leclanché cells which powered their bell or telephone systems; Dunham Massey, Greater Manchester, has two large cupboards in the attic containing dozens of cells which once powered

Fig 7.13
'Oh ah, let 'em ring again.'
Illustration by George Cruickshank from The Greatest Plague of Life: or The Adventures of a Lady in Search of a Good Servant, *by A and H Mayhew (1847).*

servant, besides, it is an aggravating thing to those who ring twice or thrice without being answered.[19]

Henry and Augustus Mayhew seem to have agreed that getting bells answered was a problem (Fig 7.13)![20]

Electric bells

Electric bells were introduced into country houses from the 1860s onwards, often well before the means of generating electricity for lighting was installed. An article in *Building News* in December 1862 claimed, 'The latest introduction of scientific application to our dwelling houses is the new system of ringing bells by electricity.'[21] As one might expect from his later involvement with electric lighting (*see*

the internal telephone system and the bells (Fig 7.14). At Osborne House on the Isle of Wight, Prince Albert's fascination with science led to many technological innovations, one of the earliest bell systems being installed there, prior to Albert's death in 1861. The writing desk in the Queen's Sitting Room incorporates three electric bell pulls, which resemble organ stops. These summoned the Queen's dresser, a page and the royal couple's personal attendant, and they were certainly powered by some earlier type of primary battery.

The electric bell itself makes use of an electromagnet linked to an armature with a small hammer. When an electrical circuit is created, the electromagnet causes the armature to move towards it and the attached hammer to strike the bell. The third element in the electrical bell system is the bell push, which forces two springs together inside the bell housing and so closes the electric circuit. The reasons for the popularity of electric bells over sprung bells may just have been fashion, but the wiring for mechanical bells required regular maintenance: Mrs Beeton's instructions for the annual spring-cleaning included the adjustment of bell wires and the tightening of bell handles, should this be necessary.[25] The space required for electrical wiring was far less than that for sprung bells, and the system was more flexible, no longer requiring that the wires should be kept in tension. Nevertheless, many houses retained their sprung bells and just installed the electrical systems alongside them, and so elements of both can often be seen.

The heart of an electrical bell system as far as the servants were concerned was the indicator or annunciator board, a shallow, glass-fronted case displaying a number of small apertures corresponding to the number of rooms from which calls could be received and each named accordingly (Fig 7.15). When a call was received, a bell close to the board rang and a coloured square or disc behind the aperture moved or fell, thus indicating where a servant was required. The discs were each activated by an electromagnet when the relevant bell push was depressed, and then had to be reset. Obviously, servants still had to hurry to the board to see where they were required, but it was easier to duplicate these boards in a number of locations than to spread the sprung bells around the house. However, Stevenson made an interesting point about an electric bell system in 1880, when he said, 'It has the disadvantage that with it

Fig 7.14
Leclanché cells and other dry batteries which once powered the bells and telephones, in an attic cupboard at Dunham Massey.

there is no power of giving expression to a ring, as of quietness or urgency, and as the same bell always rings, the servant cannot, as in ordinary bells, tell by the tone the room from which the ring proceeds, consequently someone must always be stationed beside the indicator.'[26] Whether this always happened is doubtful, but

Fig 7.15
An electric bell indicator board by the visitor's staircase at Wightwick Manor, with four 'jelly mould' electric switches below.

indicator boards can often be found in the main service passage, with additional ones outside the butler's pantry or housekeeper's room so that these senior servants could make sure the bells were answered. At Dunham Massey, for example, where the system was installed by the electric lighting specialists Drake and Gorham, there are indicator boards in the service corridor, steward's room, butler's pantry and in the attic corridor near the housemaids' bedrooms, a pattern repeated in many other houses (Fig 7.16). Equally, when the firm of Belshaw installed electricity at Brodsworth Hall in 1913, a large indicator board was placed opposite the butler's pantry. This has discs for 30 bells, including the main downstairs rooms as well as some of the principal bedrooms. A further board, containing the connections for the remaining bedrooms, was placed in the upper servants' corridor and therefore accessed by the housekeeper and female staff. It was clearly thought essential that one of the senior servants should have rapid access to the indicator boards to ensure that bells were answered quickly by the relevant servant.

Penrhyn Castle, Gwynedd, boasts a 20-indicator electrical annunciator board in the upper service passageway, with a separate manual reset for each of the three rows of indicators. This installation appears to pre-date the electricity supply to the house, as there is a battery cupboard in an adjacent part of the corridor. This

has six shelves with drop-down doors, with each shelf wide enough for at least a dozen batteries. The dates when batteries were changed are written in pencil on the insides of the doors, starting in 1917 and ending in 1927. Some but not all of the room names on the annunciator board are duplicates of the mechanical bells also present in the house, including the dining and drawing rooms and family bedrooms, but the smaller number of rooms covered by the electric bells suggests that both systems were probably in use simultaneously. A very elaborate indicator board can still be seen at Mount Stewart, County Down, made by Leo Sunderland & Co, Victoria Street, London. The board is in three parts, each of which had a bell (one of which is now missing). The indicators appear to be lights rather than the more usual electromagnetic 'flags'. The central panel is labelled 'principal rooms' and has 30 numbered lights. A blackboard below identifies the rooms for guests and has space for writing in the names of the occupants at any one time. To the left and right of the central indicator panel are two identical 20-light indicator panels, labelled 'Valets' and 'Maids'. The blackboards either side show that these were for bells in the bedrooms, so each bedroom must have had a separate bell push for maid and valet; the 'Rome' bedroom, for example, has bell pushes labelled 'M' and 'V'. From its appearance, this bell system would seem to date from the 1930s, at which point

Fig 7.16
The indicator board in a bedroom corridor at Dunham Massey, opened to show the 'flags' which moved to show where a bell had rung.

Fig 7.17 (left)
Wooden electric bell pushes
in the Bell Passage at
Tredegar House; the
electric bell system was
installed about 1921.

Lady Londonderry was regarded as one of the most important political hostesses of her time.

The bell pushes for electric bells tended to be more utilitarian than the bell levers operating sprung bells but some country houses installed more elaborate ones made of brass or porcelain to harmonise with the decor (Fig 7.17). Of all these, those at Cardiff Castle must be the most elaborate. The castle was rebuilt by the 3rd Marquess of Bute and his architect Alfred Burgess from 1872 onwards. Many of the bell pushes were made in the Bute Estate workshops and are 'conceits': one takes the form of a monkey, into whose mouth a nut is pressed to ring the bell, another is shaped like a tortoise and the bell is rung by pushing its head (Fig 7.18). For bedrooms, a pendant push, sometimes known as a pear push, was often fitted each side of the bed. The bell wire was often covered with silk, with the push at the lower end and the upper end connected to a rosette at ceiling level in which the electric circuit was made.

The wiring of electric bells provided greater flexibility than mechanical bells: as Stevenson said in 1880, 'The electric apparatus for ringing bells is sometimes adopted for houses. Its advantage is that the course of the wires being of no consequence, they can be taken into any corner or any distance, and as they do not require to be tightened they do not break.'[27] Since one bell push could operate bells in more than one location in the house, specific bells could be rung to summon particular servants more readily. For example, bells in bedrooms are sometimes labelled 'valet', or 'lady's maid', as at Mount Stewart, providing some insight into such private communications. Cragside, Northumberland, has bell pushes labelled in this way in the Owl Suite, added to the house in 1884 when Lord Armstrong was honoured by a visit from the Prince and Princess of Wales. The Rothschild house of Waddesdon, Buckingham-

Fig 7.18 (left)
An electric bell push in the
shape of a monkey, into
whose mouth a nut is
pressed to ring the bell, in
the Small Dining Room,
Cardiff Castle.

Fig 7.19 (below)
A pneumatic bell push at
Abbotsford, carved from
wood to resemble medieval
faces topped by urns.

shire, went even further by providing bedrooms with a bell push which could summon the valet, lady's maid or ring downstairs for a footman, perhaps to make up the fire, or upstairs for a maidservant, perhaps for additional hot water. These bell pushes are now in the attic store there but such examples of technology provide an insight into the inner workings of the household (*see* Fig 10.2).

Pneumatic bells were occasionally installed in country houses, making use of compressed air to push a small ball down a metal tube which then rang the relevant bell in the basement below. The most striking use of these is at Abbotsford, the home of Sir Walter Scott in the early 19th century, whose fascination with new technology is described in chapters 4 and 5 of this book. The bell pushes, topped by a metal rod for activating the compressed air, were carved from wood to resemble medieval faces topped by urns (Fig 7.19). This type of bell continued in use into the 20th century, and according to H V Margary, one of the contributors to an early 20th-century manual on modern building techniques, even then possessed many advantages over mechanical or electrical bells as they needed little maintenance.[28]

Speaking tubes

Servants still needed to go to one of the electric bell indicator boards to see in which room they were required, and then go to that room to ascertain what was wanted. The next step forwards in communications was direct speech, which would make the double journey unnecessary since requirements could be directly stated. The tubes, often a combination of metal tube and flexible pipe, were buried in walls where possible. At the transmitting end, a cylinder mouthpiece enabled the speaker to communicate with the receiving end, where a plugged-in whistle sounded to attract the attention of the intended recipient. The cylinder mouthpiece could be closed off with a plug while not in use, as can still be seen on two such mouthpieces at the top of the main stairs in Ickworth. Some speaking tube mouthpieces were also placed next to electric bells, which must have been wired to attract attention at the other end of the tube: examples can be seen at Mount Stuart on the Isle of Bute and next to the lift at Waddesdon. Here, the two bells were labelled Kitchen and Pantry, which must have saved a good deal of time in the service of food. Another Rothschild House in Buckinghamshire, Halton House, had mouthpieces labelled for communication with different floors, while at Attingham, the direction of communication of the mouthpiece in a passageway on the first floor could be changed from the floor below, the service area, to the floor above, the servants' bedrooms, by moving a lever (Fig 7.20).

Mouthpieces were available in a wide range of materials. Benham and Froud, general metal workers, for example, could provide mouthpieces in brass (lacquered, bronzed or tinned), cocoa wood, ivory, ebonite and – at a cost – silver-plated.[29] Speaking tubes enjoyed some popularity in the second half of the 19th century but lacked privacy and had a limited range (although one installed at Wightwick Manor, in the West Midlands, around 1887 connected the entrance hall of the house to the groom's room in the stable block almost 50m away). The arrival of the telephone in the 1880s, which was once again initially powered by batteries, provided the optimum solution to the problem of communicating directly within a large house and some properties had extensive internal systems, with their own switchboard, by the first decade of the 20th century, or even earlier in some cases.

Fig 7.20
A two-way speaking tube in the passageway on the first floor, Attingham Park.

Telephones

Telegraphy, sending coded messages electrically through wires, was developed in the USA and the UK in the late 1830s, where it was extensively used on the railway network. The telephone is a development of this in that the spoken word is converted into electrical impulses which are transmitted along telegraph wires to a receiver, and then converted back into speech. Alexander Graham Bell, a Scot by birth, was a teacher of the deaf and had experimented with sending musical sounds through wires by means of telegraphy. He patented his invention in the USA in 1876 and, visiting England two years later, was invited to demonstrate the new device to Queen Victoria at Osborne House on the Isle of Wight on 18 January 1878. The calls transmitted from here to London, Southampton and Cowes, as well as to the Swiss Cottage in the grounds of Osborne House, were the first long-distance calls in the UK. Unfortunately for Bell, a line from Cowes to the post room at Osborne was not installed until 1885, despite earlier promises, and the Queen never had a personal telephone.[30] However, a Bell telephone was installed at Marlborough House, Westminster, in 1878 for Alexandra, Princess of Wales, and was one of the first to be brought into practical use.[31]

Lord Salisbury at Hatfield House appears to have had very early telephones there, seemingly in 1878/9, but a reference in the archives at Hatfield implies these were French instruments and so infringed the various UK patents.[32] Lady Gwendolen Cecil, in her biography of her father, confirmed his interest in any forms of new technology, recounting:

The rather earlier invention of the telephone was only brought into practical use there [Hatfield House] at about the same time as the lights. But in 1877, soon after its publication in America but before it was on the market, Mr McLeod brought down an elementary pair of transmitters which had been manufactured under his direction by his pupils in Cooper's Hill College. Lord Salisbury at once started to send messages from one end of the house to the other, laying the connecting wires loosely over the floors of the principal rooms, to the eminent discomfort of his guests. One of them, Mr Robert Lowe, having caught his foot more than once in the wire entanglement, took a pessimistic view of the new device, and prophesied in tones of gloom that the invention would become a great bore ... The experiments were renewed a little later with a scarcely less embryonic apparatus with which, to ensure comprehension, it was found necessary to keep to familiar phrases. Visitors were startled by hearing Lord Salisbury's voice resounding oratically from selected spots within and without the house, as he reiterated with varying emphasis and expression, 'Hey diddle diddle, the cat and the fiddle; the cow jumped over the moon'.[33]

The Telephone Company Ltd was established six months after Bell's demonstration to exploit his patents and a year later Thomas Edison, another claimant to the invention of the telephone, also opened the first London exchange. After some litigation over patents, the two companies were amalgamated as the United Telephone Company in 1880. The same year, the Post Office, which had the monopoly for transmitting telegrams, won a High Court judgment that telephone calls were also a telegraphic message and so came under their monopoly. However, private telephone companies were able to continue under licence, although the Post Office also opened telephone exchanges of its own. Public exchanges then developed rapidly but most country houses, at first, purchased telephones for internal communication, particularly to outlying buildings such as the stables and generator houses. An early example of such an installation was that at Foxbury, the home of Mr H F Tiarks, at Chislehurst in Kent, almost as soon as the house was completed in the late 1870s.[34] The telephone connected the house with the stables, and the family continued to install new technologies as soon as they became available.[35] F C Allsop, in the preface to his short book on fitting telephones, published in 1892, referred to the breaking-up of the telephone monopoly which had thrown open a vast field for development, including the erection of 'private lines' in residences and stables, and stated the need for practical instruction on the working and fitting up of telephones.[36]

The essential elements of a telephone are a transmitter or microphone to speak into; a receiver or earphone which reproduces the voice of the distant person; and some sort of device, usually a bell, to announce an incoming telephone call. Early telephones were usually fitted with hand-cranked 'magneto' generators which produced a current to ring the bells of other telephones on the same line or at the exchange, as well as their own power source, usually one or more Leclanché cells, for the voice circuit. The early instruments usually installed in country houses were attached to a wall-mounted wooden board, with the magneto generator and battery housed in a wooden box attached to the board. An example of this type of telephone can be found in the corridor not far from the original butler's pantry at Chirk Castle, Wrexham, made by Peel-Connor of Manchester (Fig 7.21). This has a separate transmitter and receiver, and jack sockets for three connections to a wooden box above the instrument, labelled 'Garage', 'Power House' and 'Mr Woods', who ran the power house below the castle from 1911 to 1947.

Fig 7.21
A Peel-Connor telephone powered by dry batteries outside the butler's pantry, Chirk Castle; calls could be made to external buildings such as the garage and power house.

This instrument was probably installed early in this period and before 1923, when the Peel-Connor name was dropped as the firm had by then sold out to GEC.[37]

Manderston House has a telephone in a designated cupboard in the service corridor which served as a telephone box, complete with a shelf, although the list of numbers and the instructions which were clearly once attached to the wall are missing. This telephone also has a battery box and two bells, with the transmitter and receiver incorporated into a handset attached to the telephone on curved supports. At Castle Drogo, Devon, the telephone system was initially installed in 1915, although there are some later telephones in the house. Some of these were manufactured by Ericsson, originally a Swedish firm who formed a British-based company as early as 1903. The telephone in the electricity switch room is an Ericsson, with a transmitter and receiver on a single handset hanging beside the instrument; this connected to the power house in the valley below, where there is a similar telephone, and was used in case of an electricity failure or another problem. Another Ericsson telephone is mounted on the wall in the butler's pantry, adjacent to both the switchboard and the bell indicator.

Ericsson also introduced what is called the 'pulpit' telephone in the early part of the 20th century, of which an example survives at Brodsworth Hall, possibly dating from about 1913 (Fig 7.22).[38] This is wall-mounted, with two bells and a small sloping writing box or 'pulpit' underneath, inside which is the magneto generator, and a battery box below. Early models had a fixed microphone, but the one at Brodsworth has a handset like the telephone at Manderston. At Petworth, West Sussex, too, there is a surviving early pulpit telephone with battery box and hand-cranked magneto in the estate office. This was made by Western Electric, Antwerp, one of the European factories of the firm which had grown out of the Bell Manufacturing Company of the USA, founded very shortly after Bell's discoveries in 1882; the phone probably dates from the first decade of the 20th century. A surviving glass-fronted telephone junction box also survives on the wall near the estate office, with another in the top floor servants' corridor. The wall-mounted Bakelite telephone in the kitchen, made by the General Electric Co, Coventry, is much later in date and acts as an intercommunication system, enabling direct calls to be made to the rooms labelled beside the push

buttons as soon as the handset is lifted. It was probably installed when parts of the kitchen were modernised in the 1930s.

The telephone at Polesden Lacey, Surrey (Fig 7.23), illustrates another step forward in wall-mounted telephone design. This is a Sterling Telephone, known as an Interphone, which enabled a caller to contact as many different lines as shown on the radial line selector, in this case 10 lines. The watch receiver, or handset, had to be disconnected before a call could be made and replaced once the caller had finished. Calls made to the instrument could be taken whatever the position of the radial switch. These telephones date from the late 1920s or early 1930s, a time when Mrs Greville at Polesden Lacey was entertaining many important guests. This telephone is situated just inside the door leading from the service area into the main hall, and there is a sec-

Fig 7.22

The wall-mounted Ericsson telephone in the lower corridor at Brodsworth Hall. This was known as a 'pulpit' telephone because of the sloping writing desk below the bells. The box below would have held batteries.

ond telephone outside the butler's pantry, both convenient locations for maintaining contact between the household staff and the main rooms in the house.

Table-top telephones, with a horizontal handset incorporating both the receiver and microphone, which rested on hooks on the top of the instrument, were developed as early as the 1880s. Some were installed in country houses from the first decade of the 20th century, but fewer seem to have survived compared to wall-mounted telephones, probably because they were replaced more readily as new models became available. At Castle Drogo, there is an Ericsson table-top magneto telephone with a microphone ear trumpet in the manservant's bedroom, which probably dates from the completion of the house, since these appear in the Ericsson catalogues for 1912. The telephone in the butler's sitting room, however, has push buttons which enabled the butler to transfer calls elsewhere in the house (Fig 7.24). These handsets were only for internal communication, but handsets with a dial enabling

Fig 7.23
The Sterling Telephone in the service area at Polesden Lacey. The caller was able to move the radial line selector to choose which internal telephone to ring.

Fig 7.24
The butler's sitting room at Castle Drogo is furnished with a gramophone and a telephone handset which enabled him to transfer calls elsewhere in the house.

subscribers to connect to external exchanges were common from the 1920s. The pedestal or candlestick telephone in the State Bedroom at Kingston Lacy, Dorset, made by the General Electric Company, has a circular dial, indicating a connection to an external automatic exchange which would enable calls to be made outside the house (Fig 7.25). This model probably dates from soon after the death of Henrietta Bankes in 1923, when her son Ralph inherited the house.

Not many houses have surviving internal telephone switchboards but a very impressive example at Castle Drogo is on display to the visiting public (Fig 7.26). Dating from 1915, this is housed in the butler's pantry and contains sockets for 30 telephones. The list above the switchboard includes the farm, garage and stables, as well as rooms throughout the house, including the housekeeper's room and bedroom, the nurseries, kitchen and billiard room.

Generally, from the few examples still existing in various houses, it would seem that dealing with the telephone was the prerogative of the butler, who had replaced the steward as the controller of the house. Some butlers apparently felt hounded by internal phone systems which

Fig 7.25
A Bakelite 'candlestick'
telephone in the State
Bedroom at Kingston Lacy.

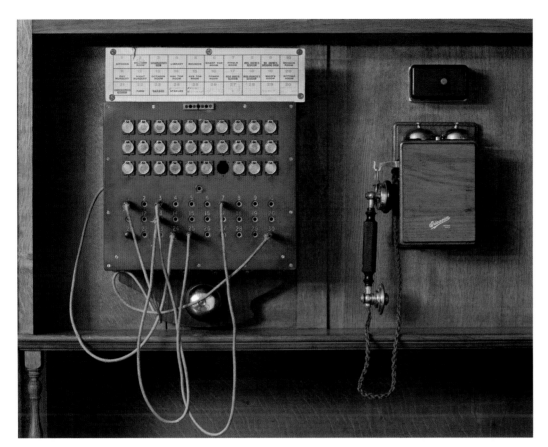

Fig 7.26
The telephone exchange in
the butler's pantry at Castle
Drogo, from which calls
could be directed to and
from the telephones within
the house.

Fig 7.27
The house communication
system in the butler's
pantry at Dunham Massey;
an electric bell indicator
board is also situated above
the telephones.

meant they were at everyone's beck and call. The old systems of wired bells did not demand an immediate response, but ringing telephones were more insistent. At Dunham Massey, for example, the butler's pantry contains two wall-mounted telephones, one with the ability to connect to five lines, below an indicator board for the electric bells and so very much a centre of communication (Fig 7.27). The butler's phone was often the only one with an external line, meaning that telegrams, too, were phoned through to this line from the local Post Office and butlers often felt the telephone never stopped ringing.[39] At Chatsworth, however, where telephones were installed as early as 1893, the Agent's list of staff in the period from 1900–12 indicates that both a telephone and a switchboard attendant were employed, although the salary paid suggests it was not a full-time post.[40] Servants were, though, permitted to make their own calls in some houses; at Castle Drogo, servants were able to make use of a telephone in a small booth in the service corridor.

The largest domestic switchboard known to survive in a country house is that at Eltham Palace, Greenwich, home of Stephen and Virginia Courtauld from 1934 to 1944. They entertained on a grand scale, the 11 family and guest bed-rooms often being fully in use. Eltham Palace was equipped in 1936 with both an external GPO telephone line and an automated internal telephone exchange for the house phones. Their letterhead gives the number as Eltham 2247. Stephen Courtauld does not seem to have been fond of the telephone, and it is significant that the only telephones connected to the external exchange appear to be those in Virginia's bedroom, boudoir and secretary's room, as well as the service corridor, not those in his own rooms.

The original Siemens internal telephone exchange still exists at Eltham and is located in a cupboard at the end of the Oak Corridor, adjacent to the entrance hall (Fig 7.28). This was a PAX 40, which had a capacity for 25 lines and operated at a voltage of from 20 to 30 volts, with a 'Transrecter' enabling it to operate directly from the mains supply.[41] This telephone exchange is a

very rare survivor and perhaps continued to operate for a time after the Army School of Education took over the premises from 1945. A number of telephones still survive within the house, although most were replaced by the army. However, the library telephone, supplied by Brothers & Co Ltd, London, is a Siemens phone and may well be the original installed in the house. Adjacent to this is a list of 18 internal telephone numbers, together with a facility to dial the fire station automatically.

The German firm of Siemens Brothers had been active in a variety of fields in the UK since the 1840s and had built a factory at Woolwich. In about 1924, in conjunction with the Post Office, they began research to overcome the defects in existing telephone handsets, which experienced considerable interference as the transmitter and receiver were too close together on the instruments. The result was the 'Neophone', which had no bell within its case and was originally used with a wall-mounted bell set. Later, bell cases were manufactured from the same materials as the telephones and could be fitted as a plinth to the base of the telephone. This telephone was originally moulded in black Bakelite, but when plastics became available shortly afterwards it could be produced in other colours. The Post Office placed large orders for these phones, which became their Telephone No. 162, and it is likely that the telephones originally installed at Eltham Palace were of this type.[42]

A further interesting feature of Eltham Palace is the telephone booth with a payphone on the left of the loggia at the far side of the entrance hall, next to the dining room (Fig 7.29). The Seely and Paget plans suggest that this was not in the original design for the house, and there was no original telephone wiring to it, but the booth was probably added very shortly afterwards.[43] As there was a limited number of external telephones in the house, this facility seems to have been provided for the many guests of Virginia and Stephen Courtauld, to enable them to make external calls – and, of course, to pay for them themselves! The coin box telephone was developed in the first decade of the 20th century but the Button 'A' and 'B' coin boxes appeared about 1925. The coin box telephone at Eltham Palace seems to have been heavily used at times: George Courtauld, a cousin of Stephen Courtauld, said in 2001 that he had been told that during the war, when many of the family were staying there for a long period, there was often a long queue for the only phone available.[44]

Fig 7.28
The Siemens PAX telephone switchboard dating from the 1930s, at Eltham Place, Greenwich.

It is likely that the sprung circular fold-down seat, cable socket, and telephone directory box at sill level (missing its middle shelf) are all original, but the telephone itself must have been removed at some stage as it was replaced by English Heritage in 2002/3.

In conclusion

Rapid communication by electronic means is an essential aspect of modern society. Studying the remains of earlier communication systems in country houses is a salutary reminder of just how recent all this is – and the evidence has all but disappeared elsewhere. Most systems were introduced to summon servants rather than for person-to-person communication, a factor that is hard to grasp today when talking is so often carried out electronically rather than face to face as it was among the families and guests in country houses. It is therefore important to record and preserve the relics of bell systems – wires, cranks, bell levers and pushes – as well as speaking tubes and early telephones, so that visitors can appreciate the way in which life in country houses worked. Country house owners, and their families and guests, would have been far less comfortable had they not been able to summon servants quickly to bring hot water up to bedrooms, make up fires or even bring drinks to gentlemen fancying a whisky while in their baths! The evidence for means of communication can be found not just in the service areas, where considerable attention has already been paid to interpretation for the visiting public, but in the state and family rooms, where the emphasis has tended to be exclusively on the furniture, paintings and decor. This chapter is a plea to recognise the importance of these relics in all areas of the house if the ways in which the households functioned are to be understood.

Fig 7.29
The coin box telephone for use by guests at Eltham Palace, Greenwich; this is a replacement for the one originally installed in the 1930s.

8

Transportation

'To ascend and descend rapidly through several flights of stairs is no trifling source of fatigue as domestic servants in some fashionable houses well know.'

Andrew Ure, 1835[1]

In the 19th century in particular, architects and country house owners gave increasing attention to the means of transporting goods around their houses and estates. As we have seen in chapters 2 and 5, the increasing size and complexity of houses and growing expectations of comfort meant that industrial quantities of coal or other fuel were required, for the boilers and furnaces in the basements and for the numerous individual fireplaces and stoves around the house, as well as in the kitchen gardens and estate yards. Frequent large house parties brought vast quantities of luggage to be moved in and out of guest rooms. The many dishes presented at lavish dinner parties had to arrive at least warm, if not hot, as seen in Chapter 6. The discerning host wanted these actions to be accomplished efficiently and, wherever possible, invisibly. The technology which facilitated this generally falls into two categories – railways for moving goods horizontally, both inside and outside the house, and lifts and hoists for moving them from floor to floor.

Railways

Coal was commonly delivered to country houses by horse and cart, usually following a discreet route around the back of the house to a service courtyard, where it could be tipped or shovelled down chutes into cellars in the basement. Examples of such coal chutes survive at many properties, including Belsay Castle (Northumberland), Belton House (Lincolnshire), Ickworth (Suffolk), Lanhydrock (Cornwall), Lyme Park (Cheshire) and Osborne House (Isle of Wight). At Castle Coole, County Fermanagh, a tunnel from the farm and service yard into the basement of the house was large enough for a horse-drawn cart to enter; turf, logs and, later, coal were stored in vaults which led directly off this tunnel (Fig 8.1). Occasionally, coal chutes were located to the front of the house, which meant that coal delivery would have been a very visible process, for example at Alnwick Castle, Northumberland, and Wrest Park, Bedfordshire. At Audley End, Essex, the chute into the coal cellar was located directly outside the main door from the house into the rear gardens, the same location used for hauling coal up to the Coal Gallery (*see* below). The means employed to deliver coal to Belvoir Castle, Leicestershire, places it in the vanguard of railway development. The castle stands on a steep rocky outcrop with the Grantham Canal, completed in 1797, looping around the bottom. In 1815, a horse-drawn tramway (Fig 8.2) was built by William

Fig 8.1
The service tunnel, Castle Coole, with bays for storing coal, wood, etc either side.

Fig 8.2
Belvoir Castle's horse-drawn
tramway, which carried coal
from the Grantham Canal to
the castle.

Jessop, of the Butterley Company of Derbyshire, from Muston Gorse Wharf for a distance of over 2km up to the castle, where a tunnel allowed the trucks to be taken directly into cellars for unloading. This tramway continued in use until 1918 and is notable for being one of the first railways to use iron fish-bellied edge rails.[2] The tunnel by which the railway entered the castle is still intact and one of the railway wagons is on display.

Harlaxton Manor, in Lincolnshire, 8km east of Belvoir Castle, also received its coal via the Grantham Canal; here, an even more elegant method was employed for getting the coal into the house. The house was built in the 1830s for the bachelor landowner Gregory Gregory, who played a prominent role in its construction (*see* Chapter 1). The house was quite technologically advanced when it was built, with an extensive warm-air heating system (*see* Chapter 5) but on Gregory Gregory's death, Harlaxton passed through a succession of increasingly distant relatives whose main properties were elsewhere, so little subsequent modernisation took place. On a site 800m south of the Grantham Canal, next to the Gregory Arms Inn on the Grantham to Melton Mowbray road, the Harlaxton estate owned a coal yard, complete with a weighbridge, which apparently served the surrounding area as well as the estate itself. From here, coal was taken by horse-drawn wagon around the back of the house to a small storage platform. Harlaxton Manor was built on a site cut into the face of a high ridge, such that ground level at the rear of

the house was roof level at the front (*see* Fig 1.25). This allowed a wooden railway to be constructed from the coal storage platform around a curved enclosed passageway into the roof space of the house (Fig 8.3). Small trucks were loaded on the storage platform and pushed into the roof space, where the coal was tipped into a number of vertical shafts which fed hoppers strategically located on different floors of the house. Alterations to the house in the 20th century have

Fig 8.3
The coal railway in the roof
space, Harlaxton Manor.
To the left of the track is one
of the shafts which fed coal
bunkers on the floors below.

Fig 8.4
The coal bunker and tunnel entrance, Chatsworth House. Man-hauled railway trucks carried coal from here to the eight boilers which heated the Great Conservatory.

Fig 8.5
The coal railway in the cellar at Tatton Park, with a turntable for changing direction.

meant that only the first section of the railway remains in the house itself; one of the coal hoppers survives at basement level in what was once the bakery.

Some country houses also required prodigious quantities of fuel to heat their garden hothouses, an extreme example being Chats-

worth, Derbyshire, where Joseph Paxton built the massive free-standing Great Conservatory in the gardens, completed in 1840 (*see* Chapter 2). As at many such estates, the gardens at Chatsworth were carefully planned to ensure that, wherever possible, the work of the gardeners remained hidden from the house, and this principle extended to coal deliveries. A sunken roadway was constructed through the gardens, with a tunnel under the Cascade, to a large below-ground storage bunker, where coal from wagons was tipped into hoppers (Fig 8.4). From there, coal was loaded into small trucks, running on iron rails, which were pushed through a tunnel a distance of 100m to the Great Conservatory, to fuel the eight boilers. These were also sited below ground, around the outside of the building, with a common chimney hidden 300m away in the woodland above the garden.

Witley Court, Worcestershire, also had a coal yard below ground level, into which coal could be tipped from carts; from here coal was transported along a railway which ran along a sunken courtyard between the Conservatory and the Stable Court and then entered the basement of the house. It then passed through a tunnel under a corner of the front gardens to reach the furnace room under the entrance hall, where there was a large warm-air heating system, similar to the Price's Apparatus described in Chapter 5. Stoke Rochford Hall, in Lincolnshire, approximately 8km south of Harlaxton, was built in the 1840s to a design by William Burn, who had been assisting Sir Gregory Gregory at Harlaxton, so it is perhaps not surprising that this, too, included a coal railway. As with Witley Court, this ran partly in an underground tunnel and was used to convey coal to a warm-air heating furnace.[3] The cellars of Osmaston Manor, Derbyshire, built 1846–9, reportedly contained a '300-foot railway with curves and turntables for carrying coal to a hydraulic lift'.[4]

The cellar corridors at Tatton Park, Cheshire, still contain the well-preserved remains of a sophisticated railway system clearly constructed to transport coal, which was delivered by cart to a below-ground bunker in the service courtyard. From here, the coal was carried in sacks or other containers on small flatbed trucks along a railway with steel rails, approximately 450mm gauge. The railway ran via a short entrance corridor to a turntable on which the trucks were turned through 90 degrees to the main track, which ran along the length of the cellar corridor (Fig 8.5). A set of points allowed the trucks to be pushed either straight on to the coal cellars or along another branch running to the foot of the lift shaft. The turntable shows the manufacturer to have been the French company Decauville, who were better known for supplying railways for mines and factories from the 1870s. Remains of another man-powered coal railway can be seen in the kitchen courtyard at Wallington, Northumberland, where trucks carried coal from a coal store to a bunker built into the wall of the kitchen, designed, like many kitchen coal bunkers, so that it could be filled from outside, thus avoiding the need for coal deliveries to pass through the kitchen.

The most extensive country house estate railway, and the only one known to have used steam locomotives, was at Eaton Hall, Cheshire, home to the Dukes of Westminster. This consisted of more than 7km of 15in (375mm) gauge track, which connected a private goods yard at Balderton, on the Great Western Railway's Shrewsbury to Cheshire line, to a yard in the service area to the north of the house, with a branch to the estate's wood yard (Fig 8.6). It was completed in 1896 and the main traffic was coal for use in the house and on the estate, including for the gasworks and electricity generating plant, but it also carried passengers between the house and the station at Balderton.[5] In South Lanarkshire, an electrically powered tramway carried passengers the 2km from Carstairs railway station to Carstairs House, whose owner, Joseph Monteith, was a prominent advocate of hydroelectricity; this possibly unique feature was built in 1888 but only operated until 1895.[6]

An unusual example of a small railway built to carry food survives at Belton House. In the

middle of the 19th century, the kitchen was moved to a building approximately 30m from the house, to which it was linked by an underground corridor (*see* Chapter 6). Attached to one wall of the corridor is a wrought-iron railway track, on which runs a small flatbed truck (Fig 8.7). It is believed that food was pushed along the corridor on this truck from the kitchen to the dumb waiter at the other end. Another food railway was installed, possibly in the 1860s, in the long basement corridor of Crewe Hall, Cheshire. Food was transported along this in a trolley, heated by a hot water jacket, from the kitchen at one end to the dumb waiter at the other; it is claimed that the flooring between the rails was made of wood blocks, to deaden the sound of the footman's footsteps.[7] Welbeck Abbey, Nottinghamshire, reportedly took delivery of a 'Dinner Wagon' in 1863, running on rails in the underground tunnel between the kitchen block and the dining room, which may have had some form of heating to keep the food warm.[8]

Fig 8.6
The narrow-gauge steam railway at Eaton Hall.

Fig 8.7
The railway in an underground passageway, at Belton House, for transporting food from the kitchen to the dumb waiter, which served the dining room.

Hoists and lifts

Manual hoists were often used to move coal, water and heavy objects between different levels. These generally consisted of a pulley mounted on a pivoted arm, with a rope wound by a windlass. At Audley End, a manual winch was installed in the top-floor Coal Gallery to haul coal from a surprisingly public location, immediately outside the main door from the house into the rear gardens. The rope ran over a wooden roller which slotted into brackets outside one of the windows and the coal, in small sacks or buckets, was swung through the open window and tipped into storage bins (*see* Fig 3.12). A replacement crane was installed in 1824 and it is likely that the original one dates from around the time the Coal Gallery was created, in the late 18th century.[9] External hoists to lift fuel to the upper floors were also installed at Birr Castle, in County Offaly, Ireland, and at Alnwick Castle.[10] A manual hoist was used to lift barrels down to the cellars at Knightshayes Court, Devon, and the wine cellar at Castle Coole has a free-standing hoist in the centre of a semi-circular stillage, which enabled barrels to be lifted and moved. At Belton House, a pivoting manual hoist was installed at the top of the servants' staircase, with a windlass housed in an iron enclosure at the back of the landing (Fig 8.8).

This enabled heavy loads to be lifted up the stairwell from the basement to any floor, but must have been quite awkward to use, as the loads had to be man-handled over the balustrades.

Lifts and dumb waiters, in which goods were conveyed in a car or cage running in an enclosed shaft, offered a more elegant solution: not only were these less visually intrusive than hoists but the risk of damage or spillage caused by the load swinging about was avoided. Dumb waiters, sometimes known as dinner lifts or rising cupboards, were provided to enable food to be moved swiftly from the kitchen area to a servery adjacent to the dining room, and to return the used dishes; examples survive in many properties, including Dyrham Park in Gloucestershire, Castle Drogo in Devon, and Sizergh Castle in Cumbria (Fig 8.9), the latter two supplied by the noted lift manufacturer Waygood & Co of London. They were generally hand-powered, with counterweights and pulleys to minimise the effort required but, as described below, hydraulic power was occasionally employed. Calke Abbey, Derbyshire, has an unusual crank-operated example with two cages, which counterbalanced each other and must also have sped up the process of moving items to and from the kitchen to the butler's pantry above. At Cragside, Northumberland, a dumb waiter by Archibald Smith and Stevens of London connects the kitchen with the scullery below; its position in front of a window suggests that this was a later addition to the kitchen (*see* Fig 6.2).

Many early lifts were installed purely for moving luggage, coal and other heavy items, and were therefore located in the service areas of houses. Generally, these were hand-powered, and so were little more than larger versions of dumb waiters; evidence for these survive, for example, at Ickworth and Lanhydrock. The earliest devices that we would recognise as mechanically powered lifts were installed in multi-storey textile mills in the early decades of the 19th century, operated by the same steam-driven line shafting that drove the mill machinery. An example, called a 'teagle', is described in detail in Andrew Ure's 1835 *The Philosophy of Manufactures*.[11] The design is credited in part to the textile manufacturer William Strutt of Belper, who, as shown in Chapter 5, was also a notable pioneer of central heating systems. Although such devices were primarily installed for moving goods, Ure's illustration depicts the lift car occupied by passengers. Perhaps the first example of a lift designed specifically for moving people was

Fig 8.8
A hoist above the servants' staircase at Belton House.

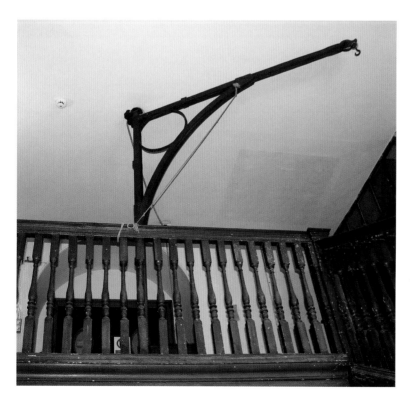

in the 'Colosseum', a circular pavilion built in London's Regent's Park in the 1820s. In 1829, *The Times* reported, 'The machinery of the ascending room is completed and visitors are now moved to the galley without the fatigue of mounting the stairs in perfect security, and by a motion which is only in the slightest possible degree discernible.'[12] It is believed that this was originally manually operated but a steam engine was installed in 1834.[13] However, country houses, warehouses and other buildings without access to steam engines initially adopted hydraulic power for their lifts.

The use of high-pressure water to power machinery was first demonstrated in the middle of the 18th century and, by the early 19th century, a number of manufacturers, including the Butterley Company of Derbyshire, were making hydraulic pumps to drain mines.[14] Early applications of hydraulic power in industry relied simply on the linear motion of the hydraulic ram and one of the challenges faced by pioneers in this field, including Joseph Bramah, who also designed early water closets (*see* Chapter 3), was to translate this into rotary motion. Early hydraulic cranes used rack-and-pinion mechanisms for both slewing the jib and winding the lifting rope, but in the 1840s William Armstrong, who soon rose to become the world's pre-eminent hydraulic engineer, developed the jigger. This consists of a pair of pulley wheels or blocks of pulleys, known as 'sheaves', separated by a hydraulic ram. A small movement of the ram, pushing the sheaves apart, is translated into a much greater movement of the rope or cable which lifts the load. Armstrong installed what may have been the first hydraulically powered lifts in the new warehouses at Liverpool docks in 1847.[15] Both forms of mechanism, as well as directly acting rams, were used in country houses.

The earliest known use of a hydraulic lift in a country house was at Osmaston Manor in Derbyshire, a large and technologically advanced mansion, built 1846–9 and demolished in 1965 (*see* Fig 1.9).[16] The owner, Francis Wright, was one of the partners in the Butterley Company, which supplied the lift, water and heating systems and much else for the house (*see* Chapter 1). This lift, like all the earliest domestic examples, was direct-acting, with the lift car raised and lowered by a hydraulic piston beneath the car. This system had two significant disadvantages: firstly, until the much later invention of telescopic rams, the cylinder had to be as long

as the length of travel of the lift, which in the case of Osmaston Manor was approximately 14m; these had to be installed below the lowest level served by the lift, which was usually the basement of the house. Secondly, unless some form of counterbalance mechanism was incorporated, the hydraulic mechanism had to lift the entire weight of the lift car and its contents, as well as the weight of the long hydraulic ram itself. This required a significant head of water, which at Osmaston Manor came from a reservoir approximately 26m above the bottom of the lift shaft. This reservoir also supplied water to the house and estate and is described in more detail in Chapter 3. This lift was designed to carry passengers, but there was a separate hydraulically powered dumb waiter which carried food from the kitchen to the dining room.[17]

The earliest known remains of hydraulically powered transport equipment in a country house are at Alnwick Castle, in the form of a dumb waiter operated by a rack-and-pinion

Fig 8.9
A Waygood manual dumb waiter in the dining room at Sizergh Castle (with the cover removed to show the mechanism).

mechanism. The ancestral home of the Percy family, later Dukes of Northumberland, since the 1300s, Alnwick was extensively modernised in the 19th century, with many technologically advanced features including central heating by steam, as described in Chapter 5. There is much evidence of the use of hydraulic power, much of which pre-dates the more well-known examples at William Armstrong's house, Cragside. In 1856 the 4th Duke of Northumberland employed a local building company to improve the castle's water supply, with reservoirs on a hillside site 3km to the west. From there, two pipes led down to the castle, one supplying drinking water and the other carrying water for hydraulic power. The work was completed by 1861. It is believed that a hydraulically powered hoist, possibly fitted to the outside of the walls, was used to bring coal from the cellars to the kitchen, and fittings within the kitchen suggest that the roasting spit was powered by a small water turbine, known as a Scotch mill (*see* Chapter 6).

However, the most remarkable survival is the large hydraulic engine installed in the basement to operate a small dumb waiter, used to carry

Fig 8.10
A mid-19th-century hydraulic engine for the dumb waiter in the basement of Alnwick Castle.

food from the basement kitchens to a corridor two floors above. This has a horizontal hydraulic cylinder which drove a rack-and-pinion mechanism (Fig 8.10). The dumb waiter, in a cupboard adjacent to the engine, was raised and lowered by a rope which ran over a pulley wheel at the top of the shaft down to another pulley at the foot of the shaft and from there to a winding drum driven by the pinion. The machinery appears massively over-engineered for the loads it had to raise. Although this innovation avoided the need to carry food up two flights of stairs, the top of the dumb waiter was still some distance from the castle's dining room. Hence, some time later in the 19th century under the 6th Duke, as part of alterations to the basement service area, a new and larger dumb waiter was installed in one of the castle's towers, close to the dining room. To reach this point at basement level, a service tunnel was dug, passing under the original castle moat. The tunnel has glass skylights and is lined with white glazed tiles to improve the light levels; it may also have had gas lighting. Food was conveyed along this tunnel, probably on trolleys, from the kitchens to a holding area at the foot of the dumb waiter, where it could be kept warm in a gas-fired hot cupboard until called for in the dining room. This later dumb waiter was operated by a direct-acting hydraulic cylinder set into the ground below basement level.

Patshull Hall, Staffordshire, had a direct-acting hydraulic goods lift and a dumb waiter, both of which survive largely intact; the former and perhaps both were supplied by Waygood & Co of London.[18] A direct-acting hydraulic goods lift was installed at Tatton Park around the 1870s, mainly to carry the coal which was transported along the railway described earlier in this chapter; this was replaced by an electric lift in 1909 (*see* below). Welbeck Abbey also had hydraulic lifts, according to Leonard Jacks' 1881 book *The Great Houses of Nottinghamshire and the County Families*, which reported: 'In order that furniture and other heavy things may be moved, the mansion is supplied with hydraulic lifts, which are constructed to work from top to bottom of the house, and in this way furniture is moved from one story [sic] to another.'[19] There was apparently also a hydraulic dumb waiter which linked with the dinner wagon railway described earlier in this chapter; it is not known whether any physical remains of these impressive systems survive.

Direct-acting hydraulic mechanisms have become popular again in recent years for glass

hydraulic cranes in the mid-19th century, installed a hydraulic lift of this type at Cragside some time around the 1870s. It had a short horizontal jigger, with triple sheaves, which can still be seen under the floor of the basement. Many lifts were installed in country houses and other buildings from 1868 by Waygood & Co, with vertical hydraulic cylinders alongside the lift shaft which operated a moving sheave in a similar manner to the conventional jigger. One was installed at Tyntesfield, near Bristol, probably in the 1870s; the lift car has been removed but the hydraulic cylinder and control mechanism survive in the Inner Yard and the shaft is now used for the modern lift.[20] There are some visible remains of a small jigger-operated hydraulic lift inserted into an existing stairwell at Wimpole Hall, Cambridgeshire; it is believed that this was also supplied by Waygood and was used by a disabled member of the family. A lift was installed by Clark, Bunnett & Co of London at Dunham Massey, Greater Manchester, in the first decade of the 20th century, with its vertical cylinder and

Fig 8.11
A lift door and hydraulic control valve in the basement at Sudbury Hall.

Fig 8.12
Direct-acting (left) and jigger-operated (right) hydraulic lift mechanisms. From The Architects' Compendium and Catalogue, *John Sears, 1907.*

lifts in commercial buildings, albeit with telescopic cylinders, because they avoid visually intrusive winding machinery. By suspending the lift car by a rope or chain passing over a pulley at the top of the shaft, counterweights could be used to reduce the hydraulic power required to move the lift. This solution was employed at Sudbury Hall, Derbyshire, where the available head of water for the hydraulic system was little more than 10m, but even with a counterbalance, the lift was restricted to a maximum load of 200kg. This lift, possibly supplied by Waygood & Co, was installed in the 1870s and had a direct-acting cylinder below basement floor level, parts of which still survive, along with the lift shaft, the air expansion vessel and other components of the hydraulic water supply system and the control valve, which was operated by means of a rope passing up the shaft (Fig 8.11).

Adoption of the jigger mechanism, first used for hydraulic cranes, brought a number of advantages to domestic lifts: the multiplying effect of the sheaves meant that the hydraulic cylinders could be shorter and use less water, and the cylinders could be installed above ground rather than having to set in deep holes below the level of the bottom of the lift shaft. This principle is illustrated in Figure 8.12. However, jiggers required a higher water pressure to lift a given load than a direct-acting cylinder. William Armstrong, whose company was a leading supplier of

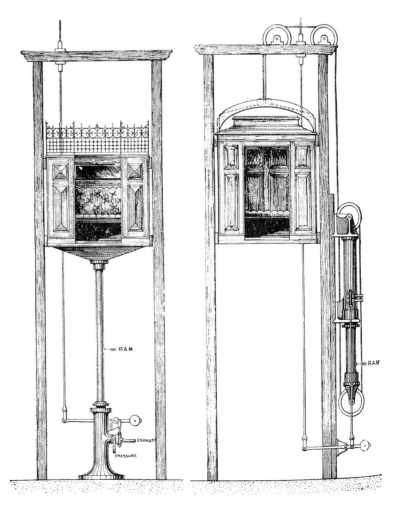

jigger mechanism located within the lift shaft itself (Fig 8.13).[21] It survives largely intact, complete with its original lift car.

By the end of the 19th century hydraulic lifts were becoming common in the luxury hotels and mansion flats which were springing up in London, where there were systems of piped hydraulic power supplied from central stations. They were normally operated by uniformed attendants. The safe operation of these early hydraulic lifts must have required some practice: the attendant pulled on a rope which operated the water control valve several floors below. There was no mechanism to ensure that the lift car stopped level with the floor of the building and the lift car doors had no safety interlocks to ensure that they could be opened only when the car was in the correct position. Electrically powered lifts overcame these shortcomings; these were pioneered in the United States in the 1880s, with Otis Brothers soon becoming the dominant manufacturer. The first electric lift in Britain was demonstrated by Way-

good & Co in 1890 at the Crystal Palace.[22] Most used electric motors to drive cable drums but some used a motor-driven screw mechanism to move the sheaves of a jigger (Fig 8.14).[23]

One of the first electric lifts shipped by Otis Brothers to the UK was installed at Waddesdon, Buckinghamshire, in 1898. The original lift car was later removed and is now on display in the Generator House (Fig 8.15). A number of British companies offered electric lifts by the 1890s, including Waygood & Co. Drake and Gorham, who built a large steam-powered electricity generator plant at Callaly Castle, Northumberland, in 1892, reportedly installed an electric lift 'for passenger and luggage purposes, powered by a 6hp motor' the following year.[24] Waygood & Co supplied an electric passenger lift for Mount Stuart, on the Isle of Bute, in the 1890s; the mechanism has been replaced but the original lift car is still in use today. Waygood were eventually taken over by Otis in 1913, although they had been collaborating on and off for some years before that:

Fig 8.13
Part of a vertical hydraulic jigger mechanism in the lift shaft, Dunham Massey.

Fig 8.14
A Waygood electrical lift. From The Architects' Compendium and Catalogue, *John Sears, 1907.*

and time. However, many of the examples described above, including Sudbury, Tatton Park and Dunham Massey, were located on the 'family' side of the door into the service area, suggesting that they might have been intended for use by passengers as well as goods, although the difficulty in controlling hydraulic lifts may have deterred family members and their guests from using them without the aid of a servant. It is certainly likely that later electric lifts, even when located in the servants' area of the house, were used by family members and guests who could not easily use the stairs. One example of this is Halton House, Buckinghamshire, where an electrically powered lift was installed in the well of the servants' staircase around the start of the 20th century; the metal balustrades were cut and hinged to form gates (Fig 8.16). The specification for the 1909 replacement electric lift at Tatton Park includes 'a hinged floor ... to fold up and protect the panelling when carrying coals or luggage', showing it was designed for both passengers and goods.[25]

It is possible that there was some initial prejudice against passenger lifts, associated as they were with 'bourgeois' hotels and mansion blocks, in a similar fashion to the apparent reluctance to install en-suite bathrooms. However, if such prejudices did exist, they were evidently overcome by practical considerations if members of the household became too infirm to use staircases; such

Fig 8.15
A 1898 Otis electric lift car, installed at Waddesdon in 1898 and now on display in the Generator House.

Fig 8.16
The remains of the electric lift inserted into the well of the servants' staircase, Halton House.

Kedleston Hall, Derbyshire, has an intact electric lift marked 'Waygood-Otis' and dated 1911.

If an electrical supply was already being provided for lighting, either from the mains or, as at Waddesdon and Mount Stuart, from local generation, electrical lifts could be cheaper to install and operate in country houses than hydraulic ones, because they removed the need for an expensive system to provide pressurised water. Electricity also made it easier to install devices to make the operation of the lift safer and more automatic. As a consequence, some properties, including Culzean Castle, in Ayrshire, and Tatton Park, replaced their original hydraulic lifts with electric ones in the 20th century.

The technological evolution of lifts within country houses is thus fairly easy to understand, but their precise role in the functioning of the house is sometimes less clear. As we have seen, some lifts were located within the service areas of houses and evidently intended wholly for the movement of goods, such as luggage, coal, water or other heavy items, saving the servants effort

considerations are believed to have prompted the installation of the lifts at Waddesdon and Mount Stuart. Royal endorsement may have also helped overcome any prejudices: a hydraulic passenger lift was installed in Buckingham Palace in 1881 but was removed five years later, apparently because the Queen found it unsightly. However, hand-powered passenger lifts were subsequently installed by Waygood & Co at Buckingham Palace, Balmoral Castle and Osborne House, the latter still surviving. Another electric lift installed in an existing stairwell survives at Polesden Lacey, Surrey; this was supplied by Otis around 1917. The lift linked Mrs Greville's apartments on the first floor and her study on the ground floor and it has been suggested that she used it not so much because of any infirmity on her part, but to facilitate a dramatic entrance from there into the library and drawing room.[26]

St Michael's Mount – a railway or a lift?

Like its 'twin' in Normandy, Mont St-Michel, St Michael's Mount in Cornwall was originally the site of a Benedictine monastery and the church, refectory and some other monastic buildings have been incorporated into the pre-

sent house. It has been owned by the St Aubyn family since 1660, although it only became their main residence in the 1870s, when Sir John St Aubyn commissioned his architect cousin Piers to extend and modernise the accommodation. The dramatic position of the house, on a steep rock cut off from the mainland except at low tide, presents particular challenges for access of people and goods (Fig 8.17). It is believed that an aerial ropeway was used initially in the 1870s to transport building materials from the harbour at the base of the rock to the house, but this was replaced with a unique inclined tramway, which can be regarded, technologically, as a cross between a railway and a lift. This 0.75m gauge tramway ran up the slope for a distance of 200m horizontally and 45m vertically, from the quayside to a platform at the foot of the walls of the house, where a door gave access to the cellars. Most of the track runs in a cut-and-cover tunnel but the final section at the top was bored through solid rock. It was constructed by local tin miners c 1880 and much of the equipment was supplied by mining equipment manufacturers Holman Brothers of Camborne.[27] It was originally powered by a gas engine, housed in the engine house at the foot of the slope, with the gas produced in a petrol-air gas plant supplied by Edmundson Ltd of London and Dublin (*see* Chapter 4).[28]

Fig 8.17
St Michael's Mount.

Fig 8.18
St Michael's Mount
tramway, powered by
an electric motor, in use
in 1963.

The gas engine drove a cable drum via a two-speed triple-reduction gear train, with the cable running along the roof of the tunnel, suspended by rollers, to a pulley wheel at the top of the track, from where it returned along the track bed to the truck. Evidently the gas engine, or, more likely, the gas plant, was not a success, as a local newspaper reported the delivery in 1897 of a 20hp oil engine, supplied by Merryweather & Co, which pumped water up to the house, probably for the fire hydrant system, as well as driving the tramway.[29]

In the 1920s, a petrol-driven dynamo was installed in the engine house to provide lighting for the house and the oil engine was replaced by a DC electric motor supplied by Messrs Drake and Gorham. When mains electricity arrived on the island in 1951, this was replaced with a 12hp AC electric motor, still driving the original winding mechanism (Fig 8.18). The tramway only ever carried goods such as coal and food, and luggage; it was never adapted to carry passengers, who had to walk or be taken on a pony up the steep path to the house. In 1990 the tramway was modernised and extended at the top to finish inside the house, in a former scullery, where a new winding mechanism was fitted to haul the truck, which remains the only means of transporting goods to the house today (Fig 8.19).

Vacuum cleaning systems

One other example of domestic technology which can be regarded as falling into the category of transportation systems is centralised vacuum cleaners, which consisted of a powerful vacuum pump, usually located in a basement, with pipes running throughout the building to sockets, to which a flexible hose and vacuum cleaning head could be connected. Cleaning

Fig 8.19
A modern tramway truck,
St Michael's Mount.

floors, and particularly carpets and rugs, had been a major household chore in Victorian and Edwardian houses. The British preference for open fires added to the problem, as dust was inevitably created in the daily servicing of these fires by housemaids and footmen. For centuries, the only effective solution was for carpets and rugs to be taken up and beaten out of doors. Experiments were carried out towards the end of the 19th century on mechanical methods of cleaning, and it was shown that the best way to remove dust from carpets was by suction rather than by attempting to blow the dust out, a system which was tried out in the USA with little success. In England the engineer Charles Booth patented the first successful vacuum cleaner in 1901, the first time that such a term had been used in print.[30] Steam-powered, wagon-mounted systems operated in London, and houses were cleaned from suction pipes passing into them from this device. Booth's success was guaranteed when his machine cleaned the Coronation carpet in Westminster Abbey for the coronation of King Edward VII in August 1902, something that made the invention known to the aristocracy. Booth formed a company called the Vacuum Cleaner Company, later the

British Vacuum Cleaner and Engineering Company, to exploit his patent and later also added the name of Goblin for his domestic cleaners.[31] Central vacuum cleaning systems were particularly popular for theatres and cinemas before the Second World War – their vacuum pumps, often powered by three-phase electric motors, provided much greater suction than portable cleaners.

An early example of a domestic central vacuum cleaning system was installed in Minterne House, Dorset, during its rebuilding between 1903 and 1907.[32] The equipment was removed when the Navy took over the house in 1940 and only the pipework and flaps for the hoses remain.[33] Another installation dating from this period was at Sennowe Park, in Norfolk, built for the descendants of Thomas Cook, the travel agent. This system was made by the Dudbridge Ironworks in Gloucestershire under licence from Charles Booth. The pump survives, housed in an outbuilding, along with the hoses and cleaning tools, but there appears to be no surviving pipework or sockets for the hoses in the house. Castle Drogo, Devon, has pipework and sockets for a centralised vacuum cleaning system but it would appear that the pump was never installed. The most complete surviving example is at Eltham Palace, Greenwich; a system supplied by the British Vacuum Cleaner Company was installed as part of its construction, completed in 1936. The vacuum pump, driven by a three-phase electric induction motor, is located in the basement (Fig 8.20). A network of pipes links this to sockets in the skirtings throughout the first and second floors of the mansion; a plan shows that this once also extended to the medieval Great Hall.[34]

In conclusion

Transportation systems, whether in the form of hoists, lifts, railways or even vacuum cleaning systems, can thus be seen as distinct from most of the other innovations featured in this book in that the technology was largely adopted to reduce the burden on servants – it was only around the end of the 19th century that lifts were widely installed for the convenience of the family and their guests. A study of these systems can often reveal useful information about the functioning of the household and a common feature of many of these systems is the extent to which they facilitated the desire for the apparently 'invisible' operation of the house and the estate.

Fig 8.20
The pump for the centralised electric vacuum cleaning system in the basement at Eltham Palace.

9

Security

'It is well known to those whose profession is concerned with the safety of country mansions that as a class these buildings are not well protected against destruction by fire.'

James Compton Merryweather, 1884[1]

Even after the primary function of the dwellings of the aristocracy ceased to be defensive, securing country houses and their occupants from attack and their contents from theft remained important. New technology played a part in achieving this more discreetly than the walls and moats of the medieval period. The other major risk to the security of the country house was fire; considerable effort and expense was devoted in the 18th and 19th centuries to improving both the fire resistance of houses and the effectiveness of firefighting equipment. However, these measures have often not been enough to save properties from fire, as illustrated by the almost total destruction of Witley Court, Worcestershire, in 1937 (Fig 9.1), and several more recent examples of serious fire damage, including Florence Court, County Fermanagh (1955), Uppark, West Sussex (1989), Windsor Castle (1992) and Clandon Park, Surrey (2015). The escalating value of artworks and metals has also meant that physical security has become an increasing priority for most country house owners in recent years.

Fig 9.1
The fire which destroyed Witley Court in 1937.

Physical security

For most country houses, the first line of defence against thieves and other intruders was their position in the centre of a large area of open parkland, usually surrounded by high walls which were also intended to deter poachers. Entrances to the park were usually guarded by gate lodges, inhabited, like most of the other houses in the area, by the families of estate workers, who might be expected to keep an eye out for any strangers. The most extreme example of a secure estate boundary is undoubtedly Wollaton Hall, Nottingham, which, as explained below, displays a number of unusual security features inside and outside the house (Fig 9.2). For most houses, strong door and window locks around the ground floor of the house were often supplemented with metal grilles and shutters. A number of houses in Northern Ireland, such as Castle Coole, in County Fermanagh, have particularly robust iron window shutters, which were probably fitted because of hostile relations between the usually Anglo-Irish owners and the surrounding population. However, such precautions were rare and most country houses relied on wooden shutters on the inside of windows,

which would have posed a limited deterrent to determined intruders and were sometimes provided to screen the sun as much as to provide security. A number of manufacturers, including Joseph Bramah and Clark, Bunnett & Co, supplied retractable metal roller shutters in the early 19th century, but these were mainly used in commercial premises such as shops and banks, although they were sometimes installed in town houses to protect windows from attack.[2] Unusually, the main ground floor windows at Brodsworth Hall, Yorkshire, have external roller shutters formed from wooden battens suspended on strips of strong fabric, which can be operated by ropes and pulleys from inside the rooms. These were supplied for the 1860s rebuilding of the house by Francis & Co of London but it is not clear why such precautions were deemed necessary.[3]

Locks are known to have existed since ancient times but their use grew from the late medieval period, reflecting perhaps a growing desire for privacy as well as the increasing wealth of the population. Earlier examples were rim locks, with the mechanism housed in a metal case on the inside face of the door (Fig 9.3); mortise locks, housed within the door itself, were intro-

Fig 9.2
Lenton Lodge, the most overtly fortified of the secure gatehouses of Wollaton Hall.

duced in the 19th century. Locks were originally just one of many products offered by local blacksmiths and metalworkers but, as with many other types of consumer goods, their manufacture became a specialised industry, in Britain concentrated in the area around Wolverhampton. Technological development accelerated from the last quarter of the 18th century, with 84 patents for locks granted in Britain between 1778 and 1851; one of the most significant was awarded in 1784 to Joseph Bramah (whose development of the water closet is described in Chapter 3). He displayed the lock in his London showroom and offered a prize of 200 guineas to anyone who could open it without a key.[4] One of the most prominent Black Country locksmiths was Jeremiah Chubb, whose company went on to manufacture safes; their products can be found in many country houses, including Petworth (West Sussex), Kingston Lacy (Dorset), Upton House (Warwickshire) and Plas Newydd (Anglesey).

Silver cutlery, plate and other tableware, being portable and easily disposed of, were most attractive for thieves and when not in use these were habitually stored in a plate safe or strong room (Fig 9.4). For added security, these safes were often located adjacent to the butler's pantry. At Stokesay Court, Shropshire, the butler's bedroom was next to the plate safe and at Castle Drogo, Devon, the silver was stored in a safe off the hall-boy's bedroom. Jewellery might also be expected to have been a target for burglars but it would appear that few houses had separate safes

for these items, although Powis Castle, Powys, has a small safe in a first-floor corridor which may have been used for this purpose. Householders were forced to rely on these physical security measures as insurers were reluctant to offer cover against burglary, regarding it as 'dangerous speculation' until the early 20th century.[5]

The other most valuable contents of country houses were often deeds and leases which, until the creation of the Land Registry in the 20th century, were essential to safeguard the estate's income. These documents were frequently stored in a secure and fire-resistant muniments safe or room and, once again, Wollaton Hall provides an extreme example. Henry Willoughby, 6th Lord Middleton, inherited Wollaton in 1800 and employed Sir Jeffry Wyatville to modernise it. The house's proximity to the burgeoning industrial town of Nottingham, at that time plagued by Luddite riots, evidently prompted Willoughby to require Wyatville to incorporate elaborate security measures: in addition to the perimeter wall and fortified gatehouses mentioned above, ground floor windows have iron frames with unique catches, staircases from the basement to the family floor have iron security doors or grilles, and a stone vaulted muniments room was created below Willoughby's study, accessible only via an iron trapdoor in the study floor. Willoughby's elaborate precautions, which extended to a well-stocked armoury, proved prescient: in October 1831, rioters enraged at the defeat of the second Reform Bill attacked and burned down Colwick Hall, 10km to the east, and, much nearer, the symbolic stronghold of Nottingham Castle but were repulsed at Wollaton.[6]

Bedroom doors could be locked from the inside by a simple bolt, for added privacy and security at night, but an elegant refinement was

Fig 9.5
A remotely operated 'night bolt' in the Chinese Bedroom, Belton House.

Fig 9.6
A watchman's register, Belton House. The patrolling watchman pulled the handle to record the time of his visit.

the 'night bolt', a lock which could be opened remotely without having to rise from the bed. Intact examples survive at Osborne House on the Isle of Wight, Hampton Court Palace, Richmond upon Thames (Greater London), Belton House in Lincolnshire (Fig 9.5) and Stokesay Court. In other houses, what appears to be the remains of a third bell pull mechanism above the bed is often evidence that a night bolt once existed. Responsibility for touring the house last thing at night, to check that all external doors and windows were securely locked and that all lamps and candles were extinguished, usually rested with the butler or another senior servant, but some properties also employed a watchman to patrol the house and its immediate surroundings at night. To ensure that this servant attended to his duties diligently and did not merely doze by the fire, a form of time clock called a watchman's register was sometimes installed at a distant point on his rounds; the watchman had to turn a key or pull a lever and the mechanical device registered the time at which he did so (Fig 9.6). Although the idea of electrically operated burglary detection systems was proposed as early as 1847, automated burglar alarm systems were rarely installed in country houses before the mid-20th century, an exception being Waddesdon, Buckinghamshire.[7]

Fire prevention and firefighting

Loss of life and property through fire was common when houses were largely constructed of timber, heating and cooking were provided by open fires, and lighting came from candles and simple oil lamps. Even when the use of brick and stone in country houses became more widespread, substantial amounts of timber continued to be used in floor and roof structures. During the medieval period, the risk of fire was often reduced by locating kitchens in a separate building from the main house and this practice continued well into the 19th century; in addition to the kitchens, brewhouses and laundries were also often located in detached buildings, linked to the main house by enclosed corridors, for example at Holkham Hall, Norfolk, and Erddig, Wrexham (where the corridor is a later addition), or, in the case of Belton House, an underground tunnel (*see* Chapters 1 and 6). Separation of these facilities had the added advantage of avoiding steam and odours entering the main house. Movable guards were placed in front of open fires to prevent sparks escaping. Weston Park, Staffordshire, has a number of 'Register' type cast-iron fireplaces with built-in curved guards which rotated behind the grate when not in use, a design patented by Joshua Jowett of High Holborn, London in 1804.[8]

Renovations and other building work created an increased risk of fire, perhaps from candles and lamps lighting work in dark spaces and from blowtorches used in plumbing (see Chapter 1). A letter to The Times in 1857 describes how, just as two years of repairs and improvements under the direction of George Gilbert Scott were nearing completion, Kelham Hall, in Nottinghamshire, was affected by fire, the precise cause of which was never established: 'A small portion of the furniture, all the plate and the family portraits are preserved – everything else is destroyed … The house and furniture were insured with the Sun Fire Office to the amount of £8,000 but it will require about twice that sum to repair the damage.'[9] This account illustrates the lax approach which country house owners had to fire insurance. The Sun Fire Office was the leader in this market, with over £4.5 million of fire cover for mansions in Britain in 1840, but the premiums generally did not cover the losses for which they had to pay out.[10] Cliveden, Buckinghamshire, a house which had been rebuilt after being almost totally destroyed by fire in 1795, was undergoing renovation following its purchase by the Duke of Sutherland in 1849 when it was again struck by fire. The Illustrated London News reported:

> The accident seems to have originated in the library, where some workmen had been employed … Messengers were instantly despatched to Maidenhead and in a very short period two engines arrived, but the fire had by that time attained so great a mastery that although an ample supply of water was at hand very little effect was produced upon the conflagration.[11]

In the 19th century, some houses incorporated measures to improve the fire resistance of the structure. In a few cases, such as the Joiners' Shop at Tatton Park, Cheshire, this mirrored the technology developed in textile mills at the end of the 18th century, with cast-iron beams and brick or concrete arches. At Florence Court the kitchen was housed in the basement of a pavilion detached from the main house but at some point in the 19th century, the Earl of Enniskillen created a museum for his prized collection of fossils on the floor above and had a cast- and wrought-iron ceiling installed in the kitchen to resist the spread of fire (Fig 9.7). The library at Alnwick Castle, Northumberland, has a mass concrete floor supported on cast-iron columns, a system patented by Fox and Barret in 1844 and mainly used in industrial and commercial buildings.[12] J C Loudon was a passionate advocate of fireproof construction, particularly for basements and kitchen areas, writing: 'Whoever lives in a house, the interior of which is subdivided by lath and plaster partitions, and which has hollow boarded floors, with a wooden staircase, is

Fig 9.7
The wrought-iron ceiling in the kitchen, Florence Court.

scarcely safer than if he dwelt over a mine of gunpowder.'[13]

Unsurprisingly, when Lanhydrock, Cornwall, was rebuilt following the disastrous fire of 1881 (*see* Chapter 1), its architect, Richard Coad, paid particular attention to fire precautions. The ceilings of the family rooms were constructed from 300mm-thick reinforced concrete slabs which were precast and shipped from London by the firm Dennett and Ingles, a feature which did not extend to servants' areas; roof trusses were made of iron and internal walls were taken up to roof height to prevent the horizontal spread of fire.[14] The Ladies' Wing of Stokesay Court, built 1889–92, has sheet asbestos between the ground floor ceilings and the floorboards above to slow the spread of fire, a relatively early example of the use of this material for this purpose.

Until the late 18th century, the chief means of extinguishing fires was with buckets of water or sand, and many country houses retain the leather buckets provided for this purpose. These were often marked with the family crest and hung from hooks in a service corridor or, as at Chirk Castle, Wrexham, in the servants' hall (Fig 9.8). At Audley End, Essex, however, where a 1797 inventory lists '24 fire buckets with crest', these were hung in a much more public area, a small lobby off the main entrance hall, which became known as the 'Bucket Hall'.[15] Relatively few houses, as described in Chapter 3, had a water supply above the ground floor until well into the 19th century (Audley End being a notable exception), so extinguishing a fire on an upper floor using only buckets must have been an impossible task.

Hand-powered water pumps for firefighting first appeared in the mid-17th century but the most successful early model was patented by Richard Newsham, a London button manufacturer, in 1721; this used chains running over toothed quadrants moved by levers to drive the

Fig 9.8
Leather fire buckets in the servants' hall, Chirk Castle.

pump pistons.[16] This design continued in production until the early 19th century, when improved pumps were introduced, most notably by two London engineering companies, Shand, Mason & Co and Merryweather & Son, who proceeded to dominate this market for over a century.[17] Two basic types of hand-powered pump were used in country houses during this period, the first being 'corridor pumps', small pumps mounted on wheels, which were kept inside the house. An example appears on a 1688 inventory at Burghley House, Lincolnshire, which lists 'a fire engine' as being stored in the Great Hall, along with 24 fire buckets.[18] Early 'corridor pumps', such as the Newsham-style pump at Powis Castle, Powys (Fig 9.9), had wooden bodies but most later models were made of iron. These pumps had a water tank which had to be filled with buckets, so they provided no greater supply of water than could be achieved with buckets alone, but they did allow the jet of water to be directed more accurately to the seat of the fire.

The second type of manual fire pump was essentially the same as the fire engines used by fire brigades in urban areas at the time. These were stored in an outbuilding and consisted of a much larger pump mounted on a cart, drawn by a team of men or horses; wide handles allowed the pump to be operated by a team of four, six or even more men. The greater power this produced meant that water could be sucked from a nearby pond, stream or other source, overcoming the constraint which the use of buckets created. Long leather or canvas hoses for suction and discharge were sometimes carried on a separate cart. In 1844, Merryweather & Son supplied such a fire engine to Audley End at a cost of almost £300; it is believed that this is the pump that still survives on site, along with its hose cart (Fig 9.10).[19] Examples survive at many other properties, including Lanhydrock, Chatsworth (Derbyshire), Tredegar (Newport) and Saltram (Devon), and a pump supplied by Joseph Bramah to Wentworth Castle (Yorkshire) in the early 19th century is on display at the Fire Service College in Gloucestershire.

The first steam-powered fire engine was developed in London around 1828 but they did not come into widespread use until the middle of the 19th century.[20] A few larger country houses invested in these, to protect not only the immediate estate but the nearby village as well. These engines were invariably horse-drawn and had locomotive-style boilers with small diameter

Fig 9.9
An early 19th-century wooden 'corridor pump', Powis Castle.

Fig 9.10
An 1844 Merryweather horse-drawn manual fire pump, Audley End.

tubes which could reach working pressure within about ten minutes. A Shand Mason steam fire engine survives at Tatton Park and is shown in a splendid photograph from around 1890 (Fig 9.11). Firms like Merryweather supplied everything an estate fire brigade might need, including uniforms (although it is not clear whether these were largely for 'show', as in this picture). They would also, for a fee, train the estate staff in firefighting and inspect them periodically. Such services could be viewed today as clever marketing, since the results of inspections were likely to focus on the need to update equipment. An inspection carried out by a Merryweather representative at Audley End in 1875 produced recommendations for improvements to the water supply and the purchase of new hoses, buckets, hand pumps and a steam-powered fire engine, at a total cost of over £1,000.[21] Subsequent correspondence between the company and the Audley agent shows that Lord Braybrooke considered this expenditure excessive; a steam-powered fire engine was sent to Audley on trial but this was returned and there is no evidence that most of the other recommendations were followed up.

James Compton Merryweather, son of the founder of Merryweather & Son, wrote a book in 1884 entitled *Fire Protection in Mansions*; this contained graphic details of fires in many country houses, showing how, in many cases, total destruction and loss of life was prevented by the use of efficient, modern firefighting equipment, which was clearly designed to encourage house owners to purchase his company's products and services. His book is full of examples of what he regards as good practice, for example, at Eastnor Castle, Herefordshire:

> A fire brigade has been formed among the men employed upon the estate; the engine examined and the men taken out for drill, I believe, on the first Monday in every month; the members of the brigade usually practise for two hours every drill day, during which time they go through all the different evolutions, so that each member is thoroughly up in every part of the work.[22]

Another influential figure who promoted fire precautions in country houses was Captain Eyre Massey Shaw, a former soldier from County Cork, who was appointed Superintendent of the Metropolitan Fire Brigade in 1861 and did much to enhance its reputation.[23] Firefighting held a fascination with some members of the aristocracy, including the Prince of Wales, which gave Shaw an entry into the most fashionable circles.

Fig 9.11
An 1890 Shand, Mason & Co steam-powered fire appliance and the estate fire brigade, Tatton Park.

He was invited to advise on fire precautions at several grand houses, including Trentham Hall, Staffordshire, and Sandringham, Norfolk, and in 1883 he was summoned by Queen Victoria to report on the arrangements at Osborne House.[24] His recommendations evidently included the purchase of a steam-powered fire pump as a brick building to house this was constructed in 1885; this building survives, along with a wooden shed dating from around 1865 which housed the manual fire pump. Shaw's fame was celebrated in song in Gilbert and Sullivan's *Iolanthe*:

> Oh, Captain Shaw! Type of true love kept under!
> Could thy brigade, with cold cascade, quench my great love, I wonder!

Improvements in water supply technology from around the start of the 19th century (*see* Chapter 3) led to the installation of dedicated fire hydrants around the periphery of many country houses; usually these connected to the main which supplied water to the house but in some instances these had a separate supply from a tank located in a tower or on nearby high ground, maintaining an elevated pressure which improved the reach of the fire pumps. Later in the 19th century it became common to install large diameter vertical pipes known as 'fire risers' throughout the house, often running up the service staircases. These had a tap at each level to which a hose, stored on a nearby reel, could be attached. Originally, such risers were kept permanently supplied with water under pressure and in some instances were fitted with gauges to indicate the pressure (Fig 9.12). At some point, however, perhaps due to problems of leakage, it became normal for these to be left empty ('dry risers'), so a valve at the bottom of the riser would have to be opened before the fire hoses could be used. Lyme Park, Cheshire, has an unusual pump house, dated 1902, on the edge of its lake; this contained an electrically powered pump which sent water from the lake either to fixed fire hydrants around the house or to fire hoses.

While designs for fire extinguishers were patented in the 18th century, the earliest successful example was demonstrated by the Norfolk inventor George William Manby in 1816.[25] This relied on compressed air to propel an aqueous solution from a canister but later examples, based on a design produced by two Frenchmen, Carlier and Vignon, around 1865, utilised a chemical reaction to produce carbon dioxide

Fig 9.12
A fire hydrant with pressure gauge, in the entrance lobby, Culzean Castle. The fire instruction notice is headed 'In case of fire, keep cool'.

which forced water from the vessel through a hose.[26] This principle remains in use to this day and many country houses retain the distinctive black steel brackets which held the most common of these extinguishers, the conical-shaped Minimax. Other fire extinguishers used carbon tetrachloride, usually housed in a glass 'grenade' which could be thrown at the fire (Fig 9.13). The most popular example of these was the Harden Star hand grenade, originally produced

Fig 9.13
A grenade-type fire
extinguisher, in the service
corridor at Erddig.

in America, and they were often provided free by insurance companies to their policyholders.[27] This type of fire extinguisher was withdrawn in the 1960s when the health hazards associated with carbon tetrachloride became known, but examples can be found in several houses, including Baddesley Clinton in Warwickshire, Calke Abbey in Derbyshire, and Erddig.

Automated fire alarms were first proposed in 1847 and their use was advocated in James Merryweather's 1884 book; a bell sounded when an electrical circuit was completed by the expansion of mercury in a tube or the severing of a flammable cord which held open a sprung switch.[28] Such a system appears to have been installed at Waddesdon but the vast majority of house owners relied on manual detection and warning of fire until well into the 20th century. Night watchmen, who, as we have seen, were employed to guard against intruders, also performed a valuable role in preventing fires – patrolling the house to check that lamps and candles were properly extinguished, for example – and raising the alarm in the event of fire. In 1900, the lives of the three children of the Duke of Portland were saved when the night watchman on his rounds in Welbeck Abbey, Nottinghamshire, saw flames coming from the vicinity of the night nursery. The *Illustrated*

London News commented that this house was 'a residence which ought to be entirely secure from fire, if man's contrivance ever could achieve that immunity'.[29] An 1875 account described these 'contrivances' soon after they were installed:

> The noble architect has apparently taken every precaution which wealth and intellect can suggest against loss by fire. Water towers and underground cisterns around the Abbey provide ample storage for water, and nearly every room is supplied with hydrants and means of applying without delay the water thus collected in case of fire. A steam fire engine and other hand engines of the best construction, with all the most approved appliances, and a well-practised fire brigade, are always in readiness.[30]

Despite these provisions and the supplementing of the estate's own fire brigade by brigades from Worksop and Sheffield, 30 rooms at Welbeck Abbey were destroyed in the fire and the loss was estimated at £40,000.[31]

In addition to fighting and preventing the spread of fires, house owners had to consider the means of escaping from them. Often, parts of houses, especially servants' quarters, were served by only a single staircase and external fixed fire escapes were rarely installed, perhaps because they might compromise the physical security of the house. Many houses still have brackets on the wall of a yard or outbuilding which once held ladders but various ingenious devices were also marketed to allay the fears of becoming trapped in a burning building. These included rope and pulley systems, by which people could be lowered from a window, and collapsible metal ladders.[32] Around the start of the 20th century, the 'Davy Automatic Safety Fire Escape' was invented. This consisted of a harness attached to a cable reel which could be fixed to the wall inside an upstairs window, which allowed people to lower themselves slowly to the ground; examples of these survive at several houses, including Penrhyn Castle, Gwynedd, and Gawthorpe Hall, Lancashire. One disadvantage of all such devices was that they needed to be rewound after use, so would have been slow to evacuate a large number of people. More rapid but perhaps also more perilous escape could be achieved by canvas chutes down which people could slide (Fig 9.14). Many houses display evidence for such equipment in the form of strong metal eyes in the wall or floor inside a bedroom window, to which the chute could be attached. At Stokesay Court a complete chute survives in a servants' bedroom, while the master bedroom has fixings and a storage locker for a collapsible metal ladder under the window seat; the chute and the ladder are detailed in a 1905 estimate from Merryweather & Son and demonstrate a distinct hierarchy when it came fire safety.[33]

In conclusion

Measures to improve security in country houses do not generally provide the most impressive or visible examples of the use of new technology and, indeed, for many properties, fire precautions did not extend beyond the provision of buckets until well into the 20th century. However, many house owners did invest significantly in measures to prevent and fight fires, either because, like Thomas Robartes at Lanhydrock, they had first-hand experience of the damage they could cause, or perhaps because they had been alarmed by the marketing publications of companies like Merryweather. The attention given to securing the house against intruders also varied significantly: perhaps many owners considered the location of their properties, in remote rural areas, surrounded by a population which was dependant upon or otherwise subservient to the estate, was protection enough. Physical security measures such as strong shutters were certainly more widely adopted for town houses; those owners who included such features in their country houses often appear to have done so out of fear of hostility from the local population, as demonstrated at Wollaton Hall and at some houses in Ireland.

Fig 9.14
A Morris & Sons 'Toboggan' fire escape. From The Country Gentlemen's Catalogue 1894.

10

Conclusion

Fig 10.1
*A painting (1830) of
Thomas Rogers, a
carpenter at Erddig. The
inscription at the base of the
painting is by Simon Yorke,
who owned the house until
his death in 1834 and, like
many of his family, wrote
many poems about the
servants at Erddig.*

*'Even the most knowledgeable country-house
enthusiasts tend to think in terms of architects,
craftsmen or family history, but know surprisingly little about how families used the houses
which the architects and craftsmen built for them.'*
Mark Girouard, 1978[1]

This book has attempted to chart the course of
technological innovation in both the country
house and its estate from the 18th to the 20th
centuries. The research was prompted by the
growing interest of some staff in the National

Trust in the preservation of technological items
in the houses belonging to them, an idea owing
something to Mark Girouard's pioneering discussion of the subject in 1978, as referred to in
Chapter 1.[2] However, the scope of this book has
embraced not only houses owned by the National
Trust and the National Trust for Scotland but
also those run by other organisations such as
English Heritage and those in private ownership,
mostly those open to the public. Such houses had
been, and many still are, at the heart of large
estates and consideration has been given to innovations in that wider context, which in some
cases preceded developments in the house itself.

Although country houses nowadays tend to
be grouped together as 'stately homes' in the
eyes of the visiting public, they were all once
owned by different people with widely varying
interests and aspirations which are reflected not
just in the architecture, interior decoration and
furnishings of their houses but also in the ways
they, their families and their large servant households lived above and below stairs. To appreciate
the differences in country house owners, one
just has to contrast the ways in which the servants were banished to an exterior block at
Petworth, West Sussex, with the close relation-
ship the servants appear to have had with the
Yorke family at Erddig, Wrexham (Fig 10.1).
One thing that owners perhaps all had in common was the recognition that the primary
function of their estates was to support the
household, and so even the more conservative
landlords were likely to sanction technological
innovation that would increase their productiv-
ity. Consequently, innovations in the means of
water supply, heating and new materials such as
cast iron and glass were often first used in the
working estate, the kitchen garden or even the
pleasure grounds, with their lakes and fountains.
In some ways, of course, technological changes
were less intrusive in these contexts, whereas
new heating or lighting systems, for example,

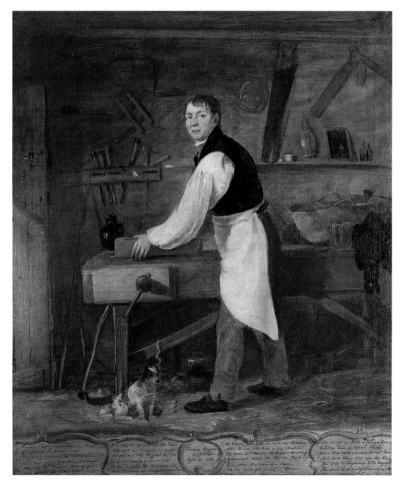

could result in considerable disruption to the fabric of ancestral homes. As has been seen throughout this book, landowners varied greatly in their attitudes towards changes of this kind within their houses, whether they were from a long line of 'old' families or 'new' men who had entered the ranks of landed society for reasons of status and prestige.

Technology transfer

Developments in industrial production, of course, played a crucial role in facilitating the growth of domestic technologies, the most important, perhaps, being the increasing availability of high-quality cast iron towards the end of the 18th century. This led to the introduction of, among other innovations, cast-iron pipes for water supplies and cast-iron stoves for kitchens, and to the use of iron in place of timber in building construction and for internal fittings in houses, garden buildings and stables. Glass, too, became easier to obtain by the mid-19th century, partly because of technical developments such as the production of sheet glass but also because of the abolition of the window tax in 1851. In some cases, too, technologies first used in industry were transferred to the domestic setting; mechanised water pumps were pioneered in mines before being adopted for domestic use from the end of the 17th century onwards, and the use of warm-air heating systems, gas lighting and hydraulic lifts followed a similar pattern. Later in the 19th century, however, it was more often in the growing numbers of large commercial buildings, such as offices, hotels and department stores, where the new forms of technology, such as steam and circulating hot water heating systems, and electric lifts, were pioneered before being introduced into domestic settings.

Equipment suppliers were often quick to seize upon the publicity value of their aristocratic, famous and wealthy clients. Astute manufacturers such as Merryweather (firefighting equipment), Blake's (ram pumps) and both Edmundson and Drake & Gorham (lighting systems) filled their sales brochures with lists of the great houses which used their products, together with testimonials from satisfied house owners. It would be interesting to learn how much this promoted their sales not just to owners of middle class villas but to those undertaking commercial buildings such as office blocks and hotels, which were growing in number and size at this time.

Perceptions of technology

Most large landowners did not remain immured on their estates throughout the year but were involved in either national or local government of one kind or another, certainly since the later years of the 17th century when Parliament had finally secured the upper hand over the monarchy. Their national and local contacts may well have familiarised them to some degree with many of the technological innovations referred to above, and some landowners adopted them out of sheer interest rather than necessarily as a means of increasing their comfort and convenience. The question of how far they were cushioned from the need to adopt labour-saving devices by the vast army of servants employed in most large houses has been debated at various points throughout this book. What is fairly certain is that increasing numbers of servants had to be housed in large service areas from the late 18th century onwards but still needed to be on call to deal with the demands of the family above stairs. The introduction of service bells made this much easier, particularly when changes in country house planning to keep the servants as invisible as possible, as well as segregated by gender, could have made finding a servant for a particular task extremely difficult. How far the servants themselves benefited from these bells, and later telephones, is doubtful since frequent calls must have increased their workload.

Water supply was often the first area to benefit from innovations, although the fact that these innovations were often first introduced to provide water for the gardens rather than for the house itself points to an innate conservatism among country house owners. Most importantly, piped water supplies paved the way for WCs and, eventually, bathrooms, and so can be seen to have significantly improved the comfort and convenience of life in the country house. Moreover, these developments must eventually have reduced the workload of servants, relieved from the burdens of carrying water and removing slops from rooms throughout the house, although bathrooms were quite a late addition to many houses.

The adoption of innovations in other areas of domestic life, including lighting, heating, cooking and communications, mostly started towards the end of the 18th century and continued throughout the 19th, with varying impact on the comfort of the household and on the work of servants. Developments in lighting can be regarded generally as successful both in reducing costs and in improving life above and below stairs, although its distribution often varied significantly throughout the house and it may well have led to longer working hours for servants. Advances in heating technology, on the other hand, did not significantly improve the levels of comfort in most houses until late in the 19th century, and sometimes much later, but once installed would have enabled more areas of the house to be used for living in and for entertaining; even then, the often massive task for housemaids and footmen of servicing numerous open fires was not entirely eliminated as the country house owner's preference for these remained strong.

Researching technological innovation

The 19th century saw a sequential development in new technologies, gas preceding electricity and warm-air central heating being succeeded by circulating hot water systems. However, as has been seen throughout this book, the dates at which country house owners chose to install these into their homes and estates did not follow a similar sequence; Sir David L Salomons might have introduced electricity into his workshops and elsewhere at Broomhill, Kent, in 1874, but Felbrigg, in Norfolk, remained obstinately un-electrified until the 1950s. Closed ranges were introduced into most large kitchens in the second half of the 19th century but seemingly old-fashioned open ranges with roasting spits were retained, or even newly installed. Parts of

Fig 10.2
Bell pushes in store in the attic, Waddesdon.

heating systems in greenhouses and forcing frames often survive from various periods, those at Calke Abbey, in Derbyshire, being a prime example. Electric bells may have been added to the communications systems from the 1860s onwards, but often only in certain areas of the house and sprung bells seem often to have continued in use alongside, while telephones seem to have been ancillary to bell systems and did not fully replace them in many cases until well into the 20th century.

Consequently, the surviving remains of technology were often adapted for other uses but, as Tim Martin and Nigel Seeley have pointed out in the National Trust's *Manual of Housekeeping*, such adaptations are often the only reason that these items survive.[3] Many light fittings, for example, were works of art in themselves and were often adapted for new fuels, chandeliers and oil lamps being transformed into electroliers (see Figs 4.25 and 5.15). Great care, then, has to be taken to work out for what fuel they were originally designed and what changes have taken place since. Cooking apparatus could also be converted to use other fuels, like the cast-iron stewing range at Burghley House, Lincolnshire, which was designed for charcoal but then changed to use gas (*see* Fig 6.25). On the other hand, many technological items became redundant when new systems were introduced but can still often be found lurking in attics and basements, like the remarkable collection of bell pushes at Waddesdon, Buckinghamshire (Fig 10.2).

Conservation, interpretation and presentation

The majority of the houses considered in this book are largely maintained by the income from the visiting public or from grants and legacies, and this is also the future for houses which are managed by the English Heritage Trust. Even those remaining in private hands are frequently being used in a very different way than in the past. Although always centres for entertaining, the number of visitors that historic houses open to the public now receive is far greater than ever before, particularly since opening hours of houses have been extended. The National Trust, charged under the 1907 Act of Parliament in which it was first incorporated with 'promoting the permanent preservation for the benefit of the nation of lands and tenements (including buildings) of beauty or historic interest', has now had to recognise that public access means it is just not possible to keep places entirely as they were. The current definition of conservation in the Trust is therefore 'careful management of change' in order to allow what is described as 'sustainable access'.[4]

In many cases, the houses taken on by national organisations such as the National Trust and English Heritage had, anyway, deteriorated from their original condition, particularly since so many of them had been requisitioned during the Second World War. It was necessary to reinstate their interiors or, as Christopher Rowell, a National Trust curator, has put it, 'to restore the spirit of a place when it has been altered in an unhistorical way'.[5] A great deal of research was undertaken to ensure that this process had a secure historical basis, but inevitably the way the houses were displayed was the decision of the staff involved, and the process by which they arrived at their decisions at the early stages of their new management was not always carefully documented. More recently, the property staff of the National Trust were encouraged to make sure that each of their houses told a particular story, and although this may have increased public interest, it has led to some manipulation of the contents of the house. Clearly, where much of the contents have been sold off to meet death duties, for example, objects have to be brought in from elsewhere to refurnish them, but considerable care has to be taken when drawing conclusions from both the existence and the location of the artefacts of early country house technology without additional verification.

Along with the massive increase in visitor numbers to historic houses open to the public, the nature of the visitors has changed as well. No longer are they mostly interested in architecture, furniture, paintings and gardens, but now also in the people who lived and worked in the houses, not just the family above stairs but the servants as well. This has led to a considerable change in the ways in which many houses are presented to the public. Although attention was paid to the service areas from the beginning in houses such as Erddig, in others these areas were either ignored or turned into staff offices, public toilets or tea rooms. The upsurge in interest in country house kitchens, pantries and laundries has led to a great deal of research on life below stairs, often by volunteers. House managers now re-route visitors to take in these areas, together with the basements, cellars and, where possible, the attics, as for example at

Belton House in Lincolnshire, Dunham Massey in Greater Manchester, and Felbrigg. In some places, this has encouraged the conservation of the remains of country house technology, as at Tredegar House, Newport, where the electric bell system now works again (see Fig 7.17), or at Audley End, Essex, with the restored Coal Gallery. In some cases, early technology has not been restored to working order but it is at least on display, like the electrical switch room at Castle Drogo, in Devon (Fig 10.3). Unfortunately, the installation of later services has occasionally disfigured the remains of domestic technology, as in the case of the important early hydraulic passenger lift at Tyntesfield, near Bristol, which survived intact but where a modern electric lift now occupies the lift shaft.[6] Some alterations to accommodate visitors are perhaps inevitable, but it is a pity if this results in the loss of features which would have been of intrinsic interest and could have contributed to telling the story of how the house operated.

Fig 10.3
The electrical switch room at Castle Drogo.

Fig 10.4
Children dressed in period
costume in the kitchen at
Attingham Park.

Fig 10.5
The replica range and
surrounding stonework
with carved inscription in
the kitchen at Gawthorpe
Hall, Lancashire.

The educational value of technological arte-facts has certainly been exploited, and visitors are now familiar with the sight of groups of schoolchildren in Victorian dress exploring the laundries or helping to cook in the kitchens (Fig 10.4). One outcome of this activity has been the replacement of items no longer in existence by replicas which help in the desired interpretation. A good example is the splendid mock range inserted into the old kitchen at Gawthorpe Hall, Lancashire, a property operated by Lancashire County Council Museum Service on behalf of the National Trust (Fig 10.5). The stonework sur-round is the original, put in by Charles Barry in the early 19th century, but the range itself was removed from the Hall during renovations in the 1960s or early 1970s. The copy was made by the Museum Service so the room could be used for school groups and to give the effect of a kitchen.[7] Another is the mock boiler at Ickworth, Suffolk, set into the wall of one of the basement rooms at Ickworth (Fig 10.6).

The Coal Gallery at Audley End is now on display to the public as a very early service area, and furnished with artefacts such as hot water cans, slipper baths and coal buckets which have been brought in, but on the basis of very thor-ough research (see Fig 3.12). Laundries and kitchens are frequently refurnished and the arte-facts on display may not all be original to the house. The kitchens at Anglesey Abbey, Cam-bridgeshire, stripped of most of their contents

following Lord Fairhaven's death in 1966, have been completely refurbished with some materials brought from elsewhere and have proved, like Audley End's Coal Gallery, to be of great interest to the public (Fig 10.7). All this is obviously meeting the need to manage change, but makes it rather difficult for the researcher trying to establish authenticity; perhaps the modern practice of documenting such developments will make this easier in the future, unlike many of the changes introduced in the 1950s and 1960s.

Self-sufficiency and energy conservation

The self-sufficiency of country house estates has been stressed throughout this book. However, the purpose of many historic houses changed dramatically in the 20th century, when their owners were no longer able to maintain them and they were taken over by the State or by charitable

Fig 10.6
The replica boiler set into the wall of the Boiler Room in the basement at Ickworth, Suffolk.

Fig 10.7
The restored kitchen at Anglesey Abbey, Cambridgeshire.

bodies like the National Trust.[8] How far can their previous self-sufficiency be maintained under this new kind of ownership and in a political climate which favours renewable sources of energy? A book about the technologies to be found in country houses cannot end without some brief discussion about recent innovations introduced to enhance the sustainability of their operation, which hark back to some of the historic features discussed in earlier chapters.

With regard to the estate yard, many workshops are still generally responsible for much of the maintenance on the estate, although in some places, such as Shugborough, in Staffordshire, or Erddig, these are themselves part of the visitor route and display blacksmiths' shops or sawmills in their historic state. Some stables, such as those at Audley End, have been brought back to life with resident horses – grooms demonstrate their feeding, grooming and exercising, to the delight of visitors. At Arlington Court, Devon, horses can sometimes be seen pulling carriages (Fig 10.8). Many kitchen gardens supply both flowers for the house and fruit and vegetables for the kitchens as they always did,

although now their products are consumed by visitors rather than the ancestral families. A number of them, such as at Audley End, for example, are run on entirely organic lines. Greenhouses are open to the public and forcing frames are full of vegetables, although no longer are there attempts to grow pineapples!

Earlier chapters in this book have shown how country houses had to take on the responsibility for both their own water and energy supplies. In the 20th century, this generally ceased as mains water and electricity supplies became available, many houses only gaining the latter for the first time during the Second World War. Now, in the 21st century, many country houses are turning back towards the self-sufficiency they had practised in earlier centuries. For example, many houses have re-introduced rainwater harvesting, albeit using modern pipework and tanks rather than the original infrastructure, the water often being used in the visitor toilet blocks, as at Chatsworth and Ilam Park, both in Derbyshire, for example. The National Trust has undertaken to reduce its use of fossil fuel by 50 per cent by 2020.[9] Biomass boilers have been installed at

Fig 10.8
Horses in loose boxes in the stables, built in 1864, at Arlington Court, Devon, where they are kept to pull the carriages in the National Trust Carriage Museum.

many sites, including around 50 National Trust properties, such as Llanerchaeron in Ceredigion, Sudbury Hall in Derbyshire, Chirk Castle, Wrexham (Fig 10.9), Scotney Castle in Kent, Tyntesfield and Castle Drogo; many more such installations are planned, and more of them are making the effort to utilise wood from the estate.[10] Various heat source pumps have also been installed to provide ancillary heat to certain areas; the marine heat source pump at Plas Newydd in Anglesey provides most of the heat for the house and outbuildings. At other properties, water source pumps have been installed in lakes, for example at Castle Howard, Yorkshire, and Packwood House, Warwickshire.[11]

However, it is in the area of hydroelectric generation that the clock has been turned back most spectacularly. Cragside, in Northumberland, is arguably the birthplace of hydroelectricity, so it is fitting that an Archimedean screw was installed in 2014, generating enough electricity to light the house again. It also provides a vivid practical demonstration for visitors of the updating of an old technology (Fig 10.10). Some other sites have reused some parts of their original hydroelectric

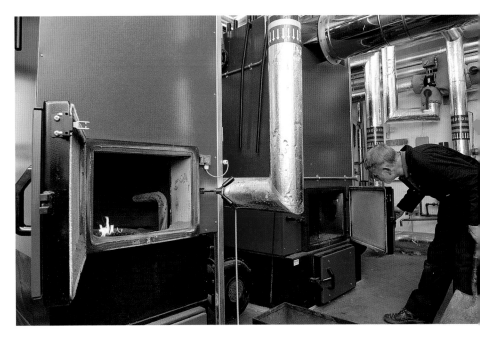

installations in their new 'green energy' schemes. Chatsworth's original turbines and dynamos, dating from 1893, survive within the Generator House but have been moved aside to make way

Fig 10.9
Checking the biomass boiler at Chirk Castle.

Fig 10.10
An Archimedean Screw at Cragside

for a modern turbine and alternator, still using the original water supply, shared with the Emperor Fountain (Fig 10.11). At Alnwick Castle, Northumberland, a Gilkes water turbine dating from the 1930s, a replacement for the original installed in 1889, has been restored and connected to a new alternator to provide power to the estate.[12] That the wheel has turned full circle with these forms of technology should not be a surprise – the surprise is, perhaps, that it has taken so long to happen. The eminent electrical engineer William Siemens wrote presciently in 1882:

> There is great room for the saving of energy in various forms. We do not depend exclusively upon coal for the power and the heat we require; we have great stores of force – the outcome of the solar radiations from day to day upon the earth, in the form of water power, of wind, tidal action. All these can, and no doubt will in time, be made useful for our purposes. I was only lately paying a visit to my friend Sir William Armstrong and there saw that he had placed one of our dynamo machines at a distance of a mile from the house under a waterfall. By this means his house was lighted by electricity. There was a brook which had run to waste from time immemorial, and now by a very simple arrangement it had been made available to light a large house entirely by electricity. How great that waste has been through the ages during which that brook has flowed, which is now utilised.[13]

Many of these new installations are on display and interpreted for the visiting public, who are increasingly interested in contemporary 'green' issues. It is difficult to miss the Archimedean screws at Cragside or at Gibson Mill in Yorkshire, while visitor attention is drawn to the biomass boiler at several houses, such as Scotney Castle and Castle Drogo. Solar panels have been discreetly positioned in several properties and 'good housekeeping' undertaken regarding draught exclusion and other heat-saving measures, something that the original shutters were meant to do. Many country houses pioneered sustainable forms of energy generation and it is entirely laudable that their current owners are not just using these again but are promoting their values beyond their boundaries to indicate what can be achieved.

Fig 10.11
The modern hydroelectric generator plant at Chatsworth, with the original 1893 equipment in the background.

Country house technology in the UK

This book has tried to develop the Country House Technology Survey set up by Nigel Seeley in the 1990s and to expand it to a wider range of properties outside the National Trust, as well as to consider these houses in the context of their estates. Country house owners have had to meet the challenge of welcoming a far wider spectrum of visitors to their properties than was the case in the past. A successful way of doing this has been to provide the public with the means of understanding how both the household and their servants lived in the house. The increased number of visits, on more days and over longer hours (certainly in the case of the National Trust) has created enormous conservation problems. Spreading out visitors by letting them see the more utilitarian areas of the houses as well as the family rooms has eased that pressure to some extent, although is has required more training of volunteer room guides, who have responded magnificently to the challenge!

In conclusion

Britain led the way in industrial innovation from the 18th century onwards and this is reflected to some extent in the development of domestic technologies, although many land-owners, despite owning mines and quarries in many instances, were often very conservative when it came to introducing these new ideas into their own houses. This is very evident, for example, in heating, where house owners in both Continental Europe and America made extensive use of heating stoves, both ceramic and cast iron, whereas British house owners remained wedded to their open fires. It would be fascinating to extend this study of country house technology to European country houses or to the larger houses in America, in order to observe the differences; American houses seem to have had bathrooms rather earlier than their British counterparts, whereas sprung bells systems do not seem to be so common in Europe. For the moment, though, this book has told the story of the development of domestic technology in houses in Britain from the point of view of what survives to be seen, and it should be read in conjunction with those books which deal more fully with the lives of servants through documentary sources. The authors hope that readers will be able to view country houses in a different way, enjoying the plasterwork, wallpapers, furniture and pictures but also realising the significance of, for example, the ornamental bell pulls on the walls of family rooms, and relating these to the rows of bells below stairs which summoned servants to their duties above stairs, and brought the two communities together.

NOTES

Preface

1 National Trust Archives, letter from Pippa Pollard, assistant to Dr Seeley, March 2000.
2 See www.hevac-heritage.org.
3 Girouard 1978.
4 Aslet 1982 and 2012.
5 Leeds City Art Galleries and Rutherford 1992.
6 Dillon 2002.
7 Sambrook 2003, 1996; *see also* Sambrook 1989 and 1999.
8 Musson 2009.
9 Sambrook and Brears 1997; Day 2009.
10 Eveleigh 1983, 2011; Grant 2013; Hopwood 2004.
11 Hardyment 1988, 1992.
12 Barnwell and Palmer 2012.
13 See, for example, Vries 1991.

1 Introduction: the background to technological change in country houses

1 Stevenson 1880, Vol II, 45.
2 Girouard 1978, 12.
3 See, for example, Sambrook 1999, Musson 2009.
4 Stevenson 1880, Vol II, 45.
5 Kerr 1864, 77.
6 Ibid 79.
7 Stevenson 1880, Vol II, 47.
8 Weaver 1911, 124.
9 Christie 2000, 29.
10 Brears 1997, 'Behind the Green Baize Door', 40.
11 Girouard 1978, 219.
12 Robinson 2012, 79.
13 Ibid 79.
14 Kerr 1864, 61.
15 Brears 1997, 'Behind the Green Baize Door'; Franklin 1981.
16 Kerr 1864, 76.
17 Holden 2012, 65–6.
18 Holden 2007, frontispiece.
19 Waterson 1980, 10.
20 Kerr 1864, 80.
21 Smith 2000, 155–6.
22 Ibid, 157–9; Adlam 2013, 9.
23 Wilson and Mackley 2000, 45.
24 Bateman 1883.
25 Girouard 1978, 8–9; Franklin 1981, 25.

26 Franklin 1981, 36–8.
27 Girouard 1978, 263.
28 Craven and Stanley 1991, 159.
29 Girouard 1979, 225.
30 Holden 2012, 60.
31 Information from Andrew McLean, former curator at Mount Stuart.
32 Bapasola 2007, 47.
33 Aslet 1982, 153; Aslet 2012, 109.
34 Palmer and West 2013, 83–90.
35 Sambrook 2003, 178–9; Gregory and Miller 2013, 38–9.
36 Morritt 1929, 168–9.
37 Hatfield House Archives: William Butterfield, *Notes on Alterations In and About Hatfield House Since 1868*, privately printed 1908, 11.
38 Bapasola 2007, 47.
39 Garnett 2001, 42.
40 Wilson and Mackley 2000, 145.
41 Ibid 120.
42 Palmer and West 2013a, 11, 53.
43 Allsop 1889, 1892.
44 Robinson 2012, 283.
45 Marshall 1996, 15, quoting from J Britton's *Architectural Antiquities of Great Britain*, Vol II, London 1809.
46 Stevenson 1880, 212–80.
47 *The Engineer*, 8 March 1907, 229–31. We owe this reference to Robert Maxwell, Consultancy Manager, National Trust, London and South-East Region.
48 Weaver, 1911.
49 Patrick 2012, 5.
50 Hiskey 2012, 23–4.
51 Robinson 2012, 325.
52 Loudon 1830, 334.
53 Oxford University Archives: Mu/3/80, Bodleian Library, brochure accompanying a tender for electric light in the Pitt Rivers Museum.
54 Merryweather 1884.
55 See Sambrook 2012, 75–92.
56 Kerr 1864, 223.
57 Gaskell 1855, 80.
58 Stevenson 1880, Vol II, 49.
59 Fletcher and Fletcher 1910, 38.
60 Phillips 1920, 9.
61 Franklin 1975, 220.
62 Gerard 1994, 160.
63 Stevenson 1880, Vol II, 280.
64 Gordon 1911, 93.
65 Franklin 1975, 107.

2 Beyond the house: technological innovation in estate buildings, parks and gardens

1 Pitt 1817, Dept 2, 87.
2 See Ward and Wilson 1971; Palmer 2012.
3 Wilkinson 1969, 306.
4 Robinson 2011, 16.
5 Thompson 1963, 156–7.
6 *See* Holt 1988, 36–53.
7 Barnwell and Giles 1997, 156–8.
8 Pitt 1817, Dept 2, 90.
9 Wilcock 1997, 21.
10 Loudon 1806, Vol II, 394.
11 *See* www.mccannhistoricbuildings.co.uk/truthaboutdovecotes (accessed 15.03.15).
12 Wade-Martins 1991, 47–8.
13 Wade-Martins 2002, 105.
14 Pitt 1817, Dept 2, 87.
15 Kerr 1864, 267; Loudon 1846, Vol 1, 359.
16 *See* Sambrook 1996.
17 Wilson and Mackley 2000, 188–9.
18 Gregory and Miller 2013, 35.
19 Marshall, Palmer and Neaverson 1992, 145–76.
20 Carson 2009, 98–9.
21 Price 1995.
22 *See* www.combemill.org (accessed 15.03.15).
23 Jacks 1881, 167.
24 Ibid 168.
25 White 1875, 147.
26 Turberville 1939, 439.
27 Waterson 1980, 146.
28 Ibid 125.
29 Worsley 2004, 2.
30 *See* www.english-heritage.org.uk/daysout/properties/bolsover-castle (accessed 15.03.15).
31 Kerr 1864, 286.
32 Worsley 2004, 248.
33 White 1875, 146.
34 *Transactions of the Thoroton Society* 1899, Vol II, report of summer excursion, 28.
35 White 1875, 146.
36 Smith 2000, 159.
37 Munby 2008.
38 Nicholson 2009.
39 Worsley 2004, 47–9.
40 Kerr 1864, 293–4.
41 Morrison and Minnis 2012, 9.
42 Minnis 2009, 2.
43 Ibid 4.

44 Smith 2010, 47.
45 Ibid 56.
46 Loudon 1835, Vol II, 719.
47 Sekers 1998, 62–66; Hiskey 2011, 43.
48 Hall 1989, 96.
49 Loudon 1835, Vol I, 577.
50 Hiskey 2011, 44.
51 Fyfe 2012, 155.
52 Grant 2013,13–14.
53 Ibid 20.
54 Loudon 1835.
55 Seligman 2008, 45.
56 Ibid 27.
57 Fyfe 2012, 171.
58 Pitt 1817, Dept 2, 91.
59 Loudon 1835, Vol I, 592–7.
60 Fyfe 2012, 172.
61 Campbell 1998, 103–6.
62 Loudon 1835, Vol II, 767–78.
63 Moss 2002, 117.
64 *The Garden Magazine*, 12 October 1872, 315.
65 Information courtesy of Robin Wright, Engineering Curator, Cragside.
66 Selignan 2008, 36.
67 Switzer 1729.
68 Roberts and Hargreaves 2003, 174.
69 Hopwood 2004, 26–7.

3 Water supply and sanitation

1 Matthews 1835, 2.
2 Colvin *et al* 1963, 549–50.
3 Lawton 2012.
4 Marshall 1996, 39–41; Girouard 1979, 248.
5 Gifford 2011.
6 Bapasola 2009, 29.
7 Morris 1982, 74.
8 www.coultershaw.co.uk (accessed Dec 2014).
9 Palmer and West 2013, 86–8, 98–102.
10 Lincolnshire Archives: BNLW 2/1/2/12 – information courtesy of Ken Hollamby.
11 Craven 1996, 79.
12 Bowden-Smith 1987.
13 Catalogue information courtesy of the Waterworks Museum, Hereford.
14 Girouard 1979, 251; Roberts and Hargreaves 2002, 174.
15 Essex Record Office: Acc A8422 11/49.
16 Fletcher and Fletcher 1910, 108–10; Bailey-Denton 1882, 180–1.
17 *The Engineer*, 8 March 1907, 230–1.
18 www.gatehouse-gazetteer.info/English%20sites/4137.html (accessed 04.07.15).
19 Hellyer 1891, 171.
20 Colvin *et al* 1963, 549–50.
21 Sambrook 1989, 24.
22 Palmer and West 2013, 85–6.
23 www.coflein.gov.uk/pdf/CPG040/ (accessed 01.01.15).
24 http://list.english-heritage.org.uk/resultsingle.aspx?uid=1183926 (accessed 26.12.14).
25 Adshead 2003, 20.
26 Gloucestershire Archives: D1610/
P58/10/4-8.
27 Glover 1830, 36.
28 Chatsworth Archives: Building Accounts, Vol IV, 446; Cavendish, 1845, 10.
29 Cavendish 1845, 115.
30 For more information on the evolution of Turkish baths, see www.victorianturkishbath.org.
31 Shifrin 2015, 247–250
32 *Glasgow Herald,* 11 May 1885. Courtesy of Andrew McLean.
33 Kerr 1864, 167.
34 Thynne 1951, 30.
35 Eveleigh 2002, 76–8.
36 Muthesius 1904, 236.
37 Eveleigh 2002, 109.
38 Muthesius 1904, 235.
39 Ibid, 237.
40 Palmer and West 2014, 11.
41 Hiskey 2012, 23–4.
42 Wright 1960, 107–8.
43 Essex Record Office: D/DBY A/43/10/ 27 July 1785.
44 Essex Record Office: D/DBY A/43/11/ 30 November 1785.
45 1787 floor plans, within 'Scrapbook' at Audley End House.
46 Sambrook 1989, 24.
47 Girouard 1978, 265.
48 Eveleigh 2011, 41.
49 Eveleigh 2002, 158.
50 Kerr 1864, 167.
51 Palmer and West 2014, 17.
52 Palmer and West 2013, 88.
53 Hiskey 2007, 7.
54 Lemmoin-Cannon 1912, 63–80.
55 Scott 1911, 85.
56 *The Engineer*, 8 March 1907, 229.
57 Singleton 2009.
58 Sambrook 1999, 198.
59 Kerr 1864, 261–3.
60 Charlton 1949, 195.
61 Sambrook 1999, 160.
62 Malcolmson 1986, 135.
63 West and Palmer 2013, 55–7.

4 Lighting and energy production

1 Lockhart 1844, 500–1.
2 For an analysis of cultural attitudes to darkness, see Ekirch 2005.
3 Swan and Swan 1968, 10.
4 Dillon 2002, 28–32. This book provides a comprehensive history of the development of all types of domestic lighting equipment. *See also* Leeds Art Galleries and Rutherford 1992.
5 Essex Record Office: D/DBy A358.
6 Evans 2011, 90.
7 Hartcup 1980, 31; Evans 2011, 67.
8 Dillon 2012.
9 O'Dea 1958, 1.
10 Shröder 1969, 59–68.
11 Wolfe 1999, 19–21.
12 Ibid 28–40.
13 Essex Record Office: D/DBy A/43/12/1785; A44/9-12; A43/3 1786-7.
14 Waterson 1980, 212–9.
15 For a detailed account of the background to the development of gas lighting, *see* Tomory 2012.
16 Falkus 1967.
17 National Gas Archive: GC:NW:ALG A/R/2.
18 Hatfield House Archives: William Butterfield, *Notes on Alterations In and About Hatfield House Since 1868*, privately printed 1908, 3.
19 Lockhart 1844, 500–1.
20 Chatsworth Archives: Accounts 1855–65.
21 Lockhart 1844, 501.
22 Lincolnshire Archives: 2TGH1/34/4/82.
23 Chatsworth Archives: Accounts 1861, 28.
24 West 2009, 86–7.
25 Keith, 1989.
26 *Journal of Gas Lighting*, 18 April 1876.
27 Carr and Gurney 1996, 35.
28 Burghley House Archives.
29 Leatham 1992, 14.
30 Information courtesy of David Good, Myra Castle.
31 O'Connor 1909.
32 Public Record Office of Northern Ireland D1071H/B/E/24/2.
33 Hiskey 2007, 17.
34 For more information on early battery developments, *see* Ayrton 1887 and Niblett 1893.
35 A good account of the development of electrical generation can be found in Bowers 1982.
36 Weightman 2011, 3–9, 39–48.
37 Hatfield House Archives: William Butterfield, *Notes on Alterations In and About Hatfield House Since 1868*, privately printed 1908, 6.
38 Broomhill Archives: 1898 Souvenir Album; McKenzie 1983, 107; Gooday 2008, 160–1.
39 Hennessey 1972, 11–20.
40 *The Engineer* 21 January 1881, 49.
41 Hatfield House Archives: Letter from Sir William Armstrong, 7 February 1881.
42 Hatfield House Archives: William Butterfield, *Notes on Alterations In and About Hatfield House Since 1868*, privately printed 1908, 6.
43 Hare 1890, 236.
44 Many accounts of such events are reported in early issues of *The Electrician* and *The Electrical Engineer*.
45 *The Electrician*, 2 June 1883.
46 Bowers, 1982, 130–8; Wilson 1957, 229–31.
47 Hunwick 2012, 124–32.
48 *The Electrician*, 2 December 1882, 66–9; for more information on Davey-Paxman steam engines, see www.paxmanhistory.org.uk.
49 Copy correspondence held at Wightwick Manor.

50 Public Record Office of Northern Ireland D1071H/B/E/24/2; Walford 1875, 28.
51 Hiskey 2007, 17–19.
52 Drake and Gorham Ltd 1905.
53 *The Engineer*, 6 March 1885, 180.
54 Gooday 2008, 78–87.
55 Information courtesy of John Beckerson, Museum of Science and Industry, Manchester.
56 Salomons 1890.
57 Holkham Hall Archives.
58 Gordon 1891.
59 Cecil 1951, 8.
60 Stokesay Court Archives.
61 Hatfield House Archives: letters from H Shillito, Clerk of Works, 1881–2.
62 Belloc: 'Newdigate Poem' (1893); Lord Finchley, from *More Peers* (1923).
63 Information courtesy of Jeri Bapasola, Archival Researcher at Blenheim; Aslet 1982, 33.
64 Knight 1906, 2–3.
65 Hiskey 2007, 20–1.
66 Hird 1911, 105–6.

5   Heating and ventilation

1   Stevenson 1880, Vol II, 219–20.
2   Combe 1842, 255.
3   Stevenson 1880, Vol II, 212–13.
4   For a detailed description of the aesthetic and technical development of the fireplace *see* Shuffrey 1912.
5   Charlton 1949, 176.
6   Franklin 1806, 228–9.
7   Eassie 1872, 181–2; Brown 1979, 165–75.
8   Cole 2010, 240–1.
9   A drawing of 1811 by Lewis Wyatt showing this arrangement survives on site at Tatton.
10 Hiskey 2007, 12
11 Loudon 1846, 1259.
12 Cooper 1958, 36.
13 *Encyclopaedia Judaica* (Gale Virtual Reference Library, accessed 13.10.14).
14 *The Gentleman's Magazine*, 64, June 1788, 562.
15 Arnott 1838; for comprehensive information on early stoves and other heating systems, see the website of the Chartered Institution of Building Services Engineers' Heritage Group: www.hevac-heritage.org; *see also* Roberts 2008.
16 Bernan 1845.
17 Robinson 2012, 15–24.
18 Stevenson 1880, 72.
19 Gloucestershire Archives: D1610/P58/4/6-7.
20 Robinson 2012, 24.
21 Wiltshire and Swindon Archives: 1325/169.
22 Menuge 1993, 52; Tann 1970, 111–15.
23 Craven 1996, 100–18.
24 Menuge 1993, 52–6.
25 Hacker 1960, 56–9.

26 Sylvester 1819.
27 Smith 2012, 44–5.
28 Essex Record Office: D/DBy A321; Loudon 1830, 108.
29 Carson 2009, 71.
30 Biographical information about G & J Haden courtesy of Frank Ferris – see www.hevac-heritage.org/victorian_engineers/haden/haden.htm (accessed 10.02.12).
31 Wiltshire and Swindon Archives: 1325/33-41 G & J Haden Order Books 1824–90.
32 Wiltshire and Swindon Archives: 1325/34 G & J Haden Order Book 1824–7.
33 Wiltshire and Swindon Archives: 1325/159 *Mansions of the nobility and gentry and public buildings warmed and ventilated* by G & J Haden.
34 Wiltshire and Swindon Archives: 1325/36 G & J Haden Order Book 1830–8.
35 *The Civil Engineer and Architect's Journal*, July 1838, 237–8.
36 Hiskey 2007, 12.
37 Wiltshire and Swindon Archives: 1325/34 G & J Haden Order Book 1824–7.
38 *Wrest Park Room Guides' Notes* (unpublished manuscript, English Heritage 2011).
39 Gloucestershire Archives: D11102 plan 181 – Ground plan of offices etc of Wrest House, Beds; undated but believed to be *c* 1904.
40 Bedfordshire and Luton Archives: BLARS X219/1 – Abstract of Expenditure for Building Wrest House 1834–40.
41 Essex Record Office: D/DBy A321 1846.
42 Historic England Archives: DP155874.
43 Palmer and West 2014, 22.
44 Billington and Roberts 1982, 112–15.
45 Ibid 146–52.
46 Ibid 115–19.
47 Richardson 1837.
48 Perkins 1840.
49 *The Times*, 20 November 1841.
50 Billington and Roberts 1982, 118.
51 Bernan 1845, 239–44.
52 Eassie 1872, 197–9.
53 Lockhart 1844, 112, 494, 500.
54 Archives of the Duke of Northumberland at Alnwick Castle: services plan with amendments dated 1899.
55 West and Palmer 2013, 8–16; Public Record Office of Northern Ireland: D/071H/B/E/24/2. Information on the use of steam at Tatton Park has been derived from plans in Cheshire Record Office collection DET/2389/.
56 West and Palmer 2013, 21–9.
57 Cheshire Record Office: DET/2389/12/d/5 and DET/2389/14.
58 Eassie 1872, 166–8.
59 Chartered Institution of Building Services Engineers 1997, 20–1.
60 Walker 1850, 107.
61 Stevenson 1880, 245.
62 Kerr 1864, 129–30.

63 Described in William Cavendish's *Methode et invention nouvelle de dresser les chevau*, 1658, quoted in Worsley 2004, 85.
64 Worsley 2004, 240–1.
65 Stevenson 1880, Vol II, 219–20.

6   Food preparation and storage

1   Beeton 1861, 25.
2   Kerr 1864, 227.
3   *See* Brears 1997, 'Behind the Green Baize Door'.
4   Kerr 1864, 228.
5   Leleux 2014, 14.
6   Kerr 1864, 214–5.
7   Beamon and Roaf 1990, 17–21.
8   *See* Weightman 2002.
9   Moss 2002, 249.
10 West and Palmer 2013, 51–3.
11 Gordon 1911, 93.
12 Hardyment 1992, 73.
13 Robinson 1983, 92–100; Du Prey 1987.
14 Brears 1997, 'The Dairy', 166–72.
15 Hewlings 2013.
16 Kerr 1864, 220.
17 Gloucestershire Archives: D1610/P58/10/4-8.
18 Brears 2009, 83.
19 Day 2009, 100.
20 Ibid 105.
21 Beeton 1861, 267.
22 Brears 1997, 'Kitchen Fireplaces and Stoves', 101.
23 Raffald 1806, frontispiece ix.
24 Brears 1997, 'Kitchen Fireplaces and Stoves', 107.
25 Loudon 1846, Vol II, 1279.
26 Sambrook 2003, 182.
27 Muthesius 1904, 96.
28 Whitbread 1992, 129.

7   Communications: bells and telephones

1   Beresford 1806, 241.
2   Girouard 1979, 27.
3   McGarry 1988, 1.9.
4   Palmer and West 2013, 89–90.
5   Loudon 1846, 282.
6   Essex Record Office: D/DBy A/44/9-12, 17 November 1786.
7   Loudon 1846, 282.
8   Essex Record Office: D/DBy Audley End Braybrooke Papers.
9   Sutcliffe 1907, 280.
10 Essex Record Office: D/DBy A/44/9-12, 17 November 1786.
11 Palmer and West 2013, 92.
12 Stevenson 1880, 278.
13 Sutcliffe 1907, 280.
14 Essex Record Office: D/DBy A/43/10/1785.
15 Yorkshire Archaeological Society Archives: DD168/2.

16 Essex Record Office: D/DBy A/43/10/1785.
17 Foulds 2011.
18 McGarry 1988, 2.6.
19 Roberts 1827, 78.
20 Mayhew and Mayhew 1847, 263.
21 McGarry 1988, 2.15, quoting from *Building News*, December 1862, 487.
22 Hatfield House Archives: William Butterfield, *Notes on Alterations In and About Hatfield House Since 1868*, privately printed 1908, 2.
23 Chatsworth Archives: Estate Cash Book for 1881, payment to Thomas Crump for the repair of electric bells. (Crump was the proprietor of a glaziers and plumbers business in Derby and had undertaken the installation of the gas plant at Chatsworth.)
24 Allsop 1889, 1.
25 Hardyment 1988, 15.
26 Stevenson,1880, 278.
27 Ibid 278.
28 Margary 1905, 195.
29 www.gracesguide.co.uk/images/5/5b/ Im1868LAX-BF.jpg (accessed 04.01.15).
30 Hann 2013, 31–2.
31 Baldwin 1925, 10.
32 Hatfield House Archives: 3M/N17, 29 February 1888, letter from the United Telegraph Company concerning the possible infringement of their patents following the installation of French-designed telephones at Hatfield House.
33 Cecil 1931, 7.
34 Baldwin 1925, 20.
35 *See* www.kemnal-road.org.uk/Pages/ Houses/Foxbury.html (accessed 01.12.14).
36 Allsop 1892, Preface.
37 www.britishtelephones.com/histpeel.htm (accessed 01.12.14).
38 www.ericssonhistory.com/products/the-telephones/Ericssons-wall-telephone-set-the-pulpit-telephone-from-1882 (accessed 20.02.15).
39 Sambrook 2012, 85.
40 Ibid, 88.
41 www.britishtelephones.com/siemensb/ pax40.htm (accessed 27.07.14).
42 www.britishtelephones.com/siemensb/ sb162.htm (accessed 27.07.14).
43 Historic England Archives: MP/ELT0085.
44 George Courtauld 'Notes, 2001' – information provided by Richard Hewlings of English Heritage.

8   Transportation

1 Ure 1835, 45.
2 Lewis 2003, 104.
3 Squires 2012, 29–31.
4 Craven and Stanley 1991, 159.
5 Eaton Estate 2008.
6 Sage 2014.
7 Gladden 2011, 62.
8 Bradbury 1989, 20.

9 Essex Record Office D/DBy A82/9 1824; *see* also Palmer and West 2013.
10 Carson 2009, 18, 67–8.
11 Ure 1835, 45–54.
12 *The Times*, 2 November 1829, 3.
13 Turvey 1993–4, 149.
14 MacNeill 1972, 13–21.
15 Ibid, 47–62; McKenzie 1983, 36–46.
16 Glynn 1853, 111–2.
17 Ibid.
18 Smith 2003.
19 Jacks 1881, 164.
20 Cooper, D A *Tyntesfield House, Wraxall, North Somerset – Report and Assessment of the Waygood Luggage/Passenger Lift c* 2008 (*see* www.cibseliftsgroup.org).
21 Irlam 2009.
22 Walker 1934, 26.
23 Gray 2002, 170–200.
24 *The Electrical Engineer*, 6 January 1893, 20.
25 Cheshire Archives: DET/2389/16/i.
26 Information courtesy of Jonathan Marsh, House and Collections Manager, Polesden Lacey.
27 Semmens 1964.
28 Public Record Office of Northern Ireland: D1071H/B/E/24/2.
29 Newby 2000.
30 Booth 1934–5.
31 The British Vacuum Cleaner Company is still trading and manufactures central vacuum cleaning systems. More information on the history of this technology can be found on their website: www.bvc.co.uk/history.html (accessed 10.10.14).
32 Aslet 2012, 109.
33 Information courtesy of Lord Digby.
34 Historic England Archives: MP/ELT0368; Palmer and West 2014, 50–7.

9   Security

1 Merryweather 1884, 5.
2 www.mileslewis.net/australian-building, section 11.06: *Finishes – Shutters* (accessed 6.10.14).
3 Yorkshire Archaeological Society Archives: DD168/2.
4 Aitken 1878, 47–53.
5 Weskill 1997, 37.
6 Smith 2012.
7 *The Builder*, 24 July 1847, 353.
8 *Burke's Annual Register* for the year 1804, 392.
9 *The Times*, 1 December 1857, 7.
10 Dickson 1960, 139.
11 *Illustrated London News*, 24 November 1849, 340.
12 www.mileslewis.net/australian-building, section 7: *Cement and Concrete* (accessed 10.09.14).
13 Loudon 1846, 876.
14 Holden 2012, 66–7.

15 Probate inventory of Sir John Griffin Griffin, taken August 1797, Audley End (Historic England Catalogue no 88073826).
16 Burgess-Wise 1977, 16.
17 Ibid 27–8.
18 Leatham 1992, 196.
19 Essex Record Office: B/BDy A321 Accounts of repairs to mansion 1825–57, April 1844.
20 Burgess-Wise 1977, 34–43.
21 Essex Record Office B/BDy A8422 Box 12.
22 Merryweather 1884, 65.
23 For an account of Shaw's life, *see* Cox 1984.
24 Cox 1984, 39–42, 91.
25 *Dictionary of National Biography* (www.oxforddnb.com, accessed 14.10.14).
26 *The London Journal*, 16 September 1865, 181.
27 This and other background information on firefighting courtesy of Mick Kernan, archivist at the Fire Service College, Moreton-in-Marsh, Gloucestershire.
28 *The Builder*, 24 July 1847, 353; Merryweather 1884, 92–3.
29 *Illustrated London News*, 13 October 1900, 530.
30 White 1875, 147.
31 *Illustrated London News*, 13 October 1900, 530.
32 Merryweather 1884, 94–5.
33 Private archives at Stokesay Court.

10   Conclusion

1 Girouard 1978, v.
2 Girouard 1978, 'Second Interlude: Early Country-House Technology', 245–266.
3 Martin and Seeley 2006, 661.
4 Staniforth 2006, 36.
5 Rowell 2006, 12.
6 See www.hevac-heritage.org/items_of_ interest/lifts/Waygood_Lift.pdf. The author of the report stressed that it was important to preserve this lift as it was a rare example of superb Victorian British engineering and had been installed before the company responsible, Waygood, was taken over by its mighty American competitor, Otis.
7 Information from Rachael Pollit, Museum Manager, Gawthorpe Hall.
8 See Mander, 1997.
9 See www.nationaltrust.org.uk/ document-1355764773127 (accessed 01.03.15).
10 See http://ntenvironmentalwork.net and www.nationaltrust.org.uk/what-we-do/ big-issues/energy-and-environment (accessed 01.03.15).
11 See www.nationaltrust.org.uk/what-we-do/big-issues/energy-and-environment/ energy-map (accessed 01.03.15).
12 Hunwick 2012.
13 Bamber 1889, 363.

HOUSE LOCATIONS

Abbotsford, Roxburghshire **P open**
A La Ronde, Devon **NT**
Alnwick Castle, Northumberland **P open**
Anglesey Abbey, Cambridgeshire **NT**
Apley Park, Shropshire **P**
Ardkinglas, Argyll **P open**
The Argory, County Armagh **NT**
Arlington Court, Devon **NT**
Armley Hall, West Yorkshire **D**
Ascott, Buckinghamshire **NT**
Aston Clinton, Buckinghamshire **D**
Attingham Park, Shropshire **NT**
Audley End, Essex **EH**

Baddesley Clinton, Warwickshire **NT**
Badminton House, Gloucestershire **P open**
Bateman's, East Sussex **NT**
Berechurch Hall, Essex **D**
Belsay Castle, Northumberland **EH**
Belton House, Lincolnshire **NT**
Belvoir Castle, Leicestershire **P open**
Beningbrough Hall, North Yorkshire **NT**
Berrington Hall, Herefordshire **NT**
Birr Castle, County Offaly, Ireland **P open**
Blaise Castle, Bristol **C**
Blenheim Palace, Oxfordshire **P open**
Blickling Hall, Norfolk **NT**
Bolsover Castle, Derbyshire **EH**
Bowood House, Wiltshire **P open**
Broadlands, Hampshire **P open**
Brodsworth Hall, South Yorkshire **EH**
Broomhill, Kent **P open**
Burghley House, Lincolnshire **P open**
Burton Agnes Hall, East Yorkshire **P open**

Calke Abbey, Derbyshire **NT**
Callaly Castle, Northumberland **P**
Canons Ashby, Northamptonshire **NT**
Canwick Hall, Lincolnshire **P**
Carisbrooke Castle, Isle of Wight **EH**
Carshalton House, Sutton
  (Greater London) **P**
Carstairs House, Lanarkshire **P**
Castell Coch, South Glamorgan **Cadw**
Castle Coole, County Fermanagh **NT**
Castle Drogo, Devon **NT**

Castle Howard, North Yorkshire **P open**
Castle Ward, County Down **NT**
Chatsworth, Derbyshire **P open**
Charlecote Park, Warwickshire **NT**
Chastleton House, Oxfordshire **NT**
Chirk Castle, Clwyd **NT**
Cliveden, Buckinghamshire **P open**
Coleshill House, Oxfordshire **NT**
Corsham Court, Wiltshire **P open**
Cotehele, Cornwall **NT**
Cragside, Northumberland **NT**
Crewe Hall, Cheshire **P open**
Culzean Castle, Ayrshire **NTS**

Didlington Hall, Norfolk **D**
Dinton Park, Wiltshire **NT**
Dodington Park, Gloucestershire **P**
Donington Park, Leicestershire **P**
Dorchester House, Westminster **D**
Dudmaston, Shropshire **NT**
Dunham Massey, Greater Manchester **NT**
Dunmore, West Lothian **P**
Dunrobin Castle, Sutherland **P open**
Dyrham Park, Gloucestershire **NT**

Eastnor Castle, Herefordshire **NT**
Eaton Hall, Cheshire **P open**
Eltham Palace, Greenwich **EH**
Erddig, Clwyd **NT**
Ewelme Down, Oxfordshire **P**
Exton Park, Rutland **P**

Felbrigg Hall, Norfolk **NT**
Foxbury, Chislehurst
  (London Borough of Bromley) **P**
Florence Court, County Fermanagh **NT**

Gawthorpe Hall, Lancashire **NT**
Gibson Mill, West Yorkshire **NT**
Gordon Castle, Moray **P open**
Greys Court, Oxfordshire **NT**
Gunton Hall, Norfolk **P**

Halton House, Buckinghamshire **P**
Ham House, Richmond upon Thames
  (Greater London) **NT**

Hampton Court Palace, Richmond upon
  Thames (Greater London) **HRP**
Hanbury Hall, Worcestershire **NT**
Harlaxton Manor, Lincolnshire **P**
Hatfield House, Hertfordshire **P open**
Henham Estate, Suffolk **P open**
Hesleyside, Northumberland **P open**
Hestercombe, Somerset **P open**
Holkham Hall, Norfolk **P open**
Houghton Hall, Norfolk **P open**

Ickworth, Suffolk **NT**
Ightham Mote, Kent **NT**
Ilam Hall, Derbyshire **YHA**
Ilam Park, Derbyshire **NT**
Ingestre, Staffordshire **P**

Kedleston Hall, Derbyshire **NT**
Kelham Hall, Nottinghamshire **P open**
Kenwood House, Hampstead
  (Greater London) **EH**
Killerton, Devon **NT**
Kingston Lacy, Dorset **NT**
Knightshayes Court, Devon **NT**
Knole, Kent **NT**
Knowsley Hall, Merseyside **P**
Kylemore Abbey, County Galway,
  Ireland **P open**

Lacock Abbey, Wiltshire **NT**
Lanhydrock, Cornwall **NT**
Llanerchaeron, Dyfed **NT**
Longleat, Wiltshire **P open**
Luscombe Castle, Devon **P**
Luton Hoo, Bedfordshire **P open**
Lyme Park, Cheshire **NT**

Marlborough House, Westminster
  (Greater London) **P**
Manderston, Scottish Borders **P open**
Mapledurham, Oxfordshire **P open**
Margam Park, West Glamorgan **C**
Marsh Court, Hampshire **P**
Mentmore Towers, Buckinghamshire **P**
Minterne House, Dorset **P open**
Mount Stuart, Isle of Bute **NT**

Mount Stewart, County Down **NT**
Myra Castle, County Down **P**

Norton Hall, South Yorkshire **P**
Nostell Priory, West Yorkshire **NT**

Oatlands Palace, Surrey **D**
Osborne House, Isle of Wight **EH**
Osmaston Manor, Derbyshire **D**
Oulton Park, Cheshire **D**

Packwood House, Warwickshire **NT**
Painshill Park, Surrey **P open**
Patshull Hall, Staffordshire **P**
Penrhyn Castle, Gwynedd **NT**
Penshurst Place, Kent **NT**
Petworth House, West Sussex **NT**
Plas Newydd, Anglesey **NT**
Polesden Lacey, Surrey **NT**
Powis Castle, Powys **NT**

Rufford Old Hall, Lancashire **NT**

St Michael's Mount, Cornwall **NT**
Saltram House, Devon **NT**
Sandringham House, Norfolk **P open**
Scotney Castle, Kent **NT**
Sennowe Park, Norfolk **P**
Shipley Hall, Derbyshire **D**
Shugborough, Staffordshire **C and NT**
Sizergh Castle, Cumbria **NT**
Soho House, West Midlands **P open**
Somerleyton Hall, Suffolk **P open**
Somerset House, Westminster
  (Greater London) **P open**
Stoke Rochford Hall, Lincolnshire **P open**

Stokesay Court, Shropshire **P open**
Stratfield Saye, Hampshire **P open**
Studley Royal, North Yorkshire **NT**
Sudbury Hall, Derbyshire **NT**
Sudeley Castle, Gloucestershire **P open**
Sunnycroft, Shropshire **NT**

Tatton Park, Cheshire **C and NT**
Temple Belwood, Lincolnshire **D**
Traquair House, Scottish Borders **P open**
Tredegar House, Gwent **NT**
Trentham Hall, Staffordshire **P**
Tring Park Mansion, Hertfordshire **P**
Tullynally Castle, County Westmeath,
  Ireland **P open**
Tyntesfield, Somerset **NT**

Uppark, West Sussex **NT**
Upton House, Warwickshire **NT**

Waddesdon Manor, Buckinghamshire **NT**
Wallington, Northumberland **NT**
Warwick Castle, Warwickshire **P open**
Welbeck Abbey, Nottinghamshire **P open**
Wentworth Castle,
  South Yorkshire **P open**
Westonbirt, Gloucestershire **P open**
Weston Park, Staffordshire **P open**
Wightwick Manor, West Midlands **NT**
Wimpole Hall, Cambridgeshire **NT**
Witley Court, Worcestershire **EH**
Woburn Abbey, Bedfordshire **P open**
Woodchester, Gloucestershire **P open**
Wollaton Hall, Nottingham **C**
Worsley New Hall, Greater Manchester **D**
Wrest Park, Bedfordshire **EH**

**Key to the management/ownership
of the above named locations at
time of publishing:**

C = Council

Cadw = Historic Environment Division of
the Welsh Government

D = Demolished

EH = English Heritage Trust

HRP = Historic Royal Palaces

NT = National Trust

NTS = National Trust for Scotland

P = Privately owned (ie not open to the
public)

P open = Privately owned and open to
the public at certain times during the year
or on specific 'open days' and operating
as either one or a combination of the
following: art gallery; B&B; country house;
gardens; hotel; museum; restaurant;
shop (farm or gift); stately home.

YHA = Youth Hostel Association

ILLUSTRATION CREDITS

Historic England would like to acknowledge the following copyright holders and thank those who have given permission for the reuse of images.

Authors (Marilyn Palmer & Ian West)

Fig 1.4; Fig 1.16; Fig 1.19; Fig 1.23; Fig 2.5; Fig 2.9; Fig 2.10; Fig 2.11; Fig 2.12; Fig 2.17; Fig 2.23; Fig 2.24; Fig 3.1; Fig 3.2; Fig 3.5; Fig 3.10; Fig 3.12; Fig 3.22; Fig 3.23; Fig 3.25; Fig 3.28; Fig 3.29; Fig 3.30; Fig 3.36; Fig 4.4; Fig 4.5; Fig 4.11; Fig 4.13; Fig 4.15; Fig 4.16; Fig 4.25; Fig 4.29; Fig 5.5; Fig 5.7; Fig 5.12; Fig 5.15; Fig 5.16; Fig 5.17; Fig 5.18; Fig 5.20; Fig 5.23; Fig 5.24; Fig 6.4; Fig 6.5; Fig 6.8; Fig 6.9; Fig 6.10; Fig 6.11; Fig 6.15; Fig 6.16; Fig 6.17; Fig 6.23; Fig 6.28; Fig 6.29; Fig 7.6; Fig 7.7; Fig 7.11; Fig 7.14; Fig 7.16; Fig 7.20; Fig 7.23; Fig 7.29; Fig 8.1; Fig 8.7; Fig 8.8; Fig 8.9; Fig 8.11; Fig 8.13; Fig 8.15; Fig 8.19; Fig 8.20; Fig 9.2; Fig 9.5; Fig 9.7; Fig 9.9; Fig 9.10; Fig 9.13; Fig 10.2; Fig 10.10

(the list below is the copyright of the authors but reproduced with permission as follows:)

reproduced with permission of the Abbotsford Trust Fig 7.19

reproduced with permission of Birmingham Museums Trust Fig 5.10

reproduced with permission of the Burghley House Preservation Trust Fig 4.10; Fig 6.14; Fig 6.25

reproduced with permission of Cardiff Castle Fig 7.18

reproduced with permission of Chatsworth Settlement Trustees Fig 2.19; Fig 2.25; Fig 8.4; Fig 10.11

reproduced with permission of Viscount Coke and the Trustees of the Holkham Estate Fig 3.38

reproduced with permission of Charles Harvey Combe's grandson, Dominic Combe Fig 2.27

reproduced with permission of David Good Fig 2.2; Fig 4.14; Fig 6.12

reproduced with permission of the Halton House Archives Fig 4.28; Fig 8.16

reproduced with permission of Harlaxton College Fig 8.3

reproduced with permission of Heligan Gardens Ltd Fig 2.21

reproduced with permission of Houghton Hall, Norfolk Fig 3.8

reproduced as a free download from Library of Congress Fig 5.2

reproduced with permission of the National Trust for Scotland Fig 5.25; Fig 9.12

reproduced with permission of the Duke of Northumberland Fig 6.13; Fig 8.10

reproduced courtesy of Lester Oram, English Heritage Trust Fig 7.28

reproduced with permission of Lord Palmer, Manderston, www.manderston.co.uk Fig 2.14; Fig 7.2

reproduced courtesy of the Marquess of Salisbury, Hatfield House Fig 4.20

reproduced courtesy of Salomons UK Ltd Fig 2.15; Fig 4.24; Fig 5.26

reproduced with permission of Stokesay Court, www.stokesaycourt.com Fig 1.10; Fig 3.32; Fig 5.19

reproduced courtesy of The Sudbury Estate Fig 4.12

reproduced with permission of Traquair Charitable Trust Fig 7.3; Fig 7.8

reproduced courtesy of University of Leicester Special Collections Fig 4.3; Fig 7.13

reproduced with permission of the Trustees of the Weston Park Foundation, www.weston-park.com Fig 2.16; Fig 6.30

reproduced with permission of Wollaton Hall, Nottingham City Council Fig 1.24; Fig 5.3

Bill Barksfield, Heritage of Industry Ltd, www.heritageofindustry.co.uk

Fig 1.7; Fig 4.6; Fig 7.21

reproduced with permission of Lord Palmer of Manderston, www.manderston.co.uk Fig 7.5

Cheshire Record Office, Cheshire Archives and Local Studies and the owner/depositor to whom copyright is reserved

Fig 5.22 DET2389/12/d/5

Country Life Picture Library

Fig 1.18 1227630; Fig 2.20 728528

Daily Herald Archive, National Media Museum, Science & Society Picture Library - all rights reserved

Fig 9.1 10249873

Gloucestershire Archives

Fig 3.15 D16610/P58/1/51

Harlaxton College

Fig 1.25

Historic England Archive

Fig 1.9 OP22037; Fig 1.11 DP070720; Fig 2.1 DP031122; Fig 2.26 DP082965; Fig 3.6 DP186302; Fig 7.22 DP028977

Illustrated London News Ltd, Mary Evans Picture Library

Fig 4.18

BIBLIOGRAPHY

Adlam, D 2013 *Tunnel Vision: The Enigmatic 5th Duke of Portland.* Worksop: The Harley Gallery

Adshead, D 2003 '"Like a Roman Sepulchre": John Soane's Design for a *Castello d'Acqua* at Wimpole, Cambridgeshire, and its Italian Origins'. Apollo, 157/494, 15

Aitken, W C 1878 'Locks', in Bevan, G P (ed) *British Manufacturing Industries: The Birmingham Trades,* 47–57. London: Edward Stanford

Allsop, F C 1889 *Practical Electric Bell Fitting: A Treatise on the Fitting-Up and Maintenance of Electric Bells and All the Necessary Apparatus.* (Facsimile edition Bibliolife, Charleston, South Carolina, 2009)

Allsop, F C 1892 *Telephones, Their Construction and Fittings: A Practical Treatise on the Fitting-Up and Maintenance of Telephones and their Auxiliary Apparatus.* (Facsimile edition Kessinger Publishing, Whitefish, Montana, 2009)

Arnott, N 1838 *On Warming and Ventilating.* London: Longman, Orme, Brown, Green and Longmans

Aslet, C 1982 *The Last Country Houses.* New Haven: Yale University Press

Aslet, C 2012 *The Edwardian Country House: A Social and Architectural History.* London: Frances Lincoln (updated edition of Aslet 1982)

Ayrton, E 1887 *Practical Electricity.* London: Cassell & Co

Bailey-Denton, E F 1882 *Handbook of House Sanitation.* London: E & F N Spon

Baldwin, F G C 1925 *The History of the Telephone in the United Kingdom.* London: Chapman Hall

Bamber, E F (ed) 1889 *The Scientific Works of C William Siemens, Vol III: Addresses, Lectures, etc.* London: John Murray

Bapasola, J 2007 *Household Matters: Domestic Service at Blenheim Palace.* Woodstock: Blenheim Palace

Bapasola, J 2009 *The Finest View in England: The Landscape and Gardens at Blenheim Palace.* Woodstock: Blenheim Palace

Barnwell, P S and Giles, C 1997 *English Farmsteads, 1750–1914.* London: RCHME

Barnwell, P S and Palmer, M (eds) 2012 *Country House Technology (Rewley House Studies in the Historic Environment 2).* Donington (Lincolnshire): Shaun Tyas

Bateman, J 1883 *The Great Landowners of Great Britain and Ireland,* 4th edn. London: Harrison; reprinted 1971 by Leicester University Press, New York

Beamon, S P and Roaf, S 1990 *The Ice-Houses of Britain.* London: Routledge

Beeton, I M 1861 *Beeton's Book of Household Management.* London: S O Beeton Publishing

Beresford, J 1806 *The Miseries of Human Life, or the Groans of Timothy Testy and Samuel Sensitive.* London: William Miller

Bernan, W (pseudonym of R S Meikleham) 1845 *On the History and Art of Warming and Ventilating Rooms and Buildings ... (2 Volumes).* London: George Bell

Billington, N S and Roberts, B M 1982 *Building Services Engineering: A Review of its Development.* Oxford: Pergamon Press

Booth, H C 1934–5 'The Invention of the Vacuum Cleaner'. *Transactions of the Newcomen Society* 15, 85–98

Bowden-Smith, R 1987 *The Water House, Houghton Hall, Norfolk (English Garden Features 1600–1900, No 1).* Woodbridge, Suffolk: Avenue Books

Bowers, B 1982 *A History of Electric Lighting and Power.* Stevenage: Peter Peregrinus

Bradbury, D J 1989 *Welbeck and the 5th Duke of Portland.* Mansfield: Wheel Publications

Brears, P 1997 'Behind the Green Baize Door', in Sambrook, P A and Brears, P (eds) *The Country House Kitchen 1650–1900.* London: National Trust Enterprises in association with Sutton Publishing Ltd (paperback edn), 30–76

Brears, P 1997 'The Dairy', in Sambrook, P A and Brears, P (eds) *The Country House Kitchen 1650–1900.* London: National Trust Enterprises in association with Sutton Publishing Ltd (paperback edn), 164–175

Brears, P 1997 'Kitchen Fireplaces and Stoves', *in* Sambrook, P A and Brears, P (eds) *The Country House Kitchen 1650–1900.* London: National Trust Enterprises in association with Sutton Publishing Ltd (paperback edn), 92–115

Brears, P 2009 'The Roast Beef of Windsor Castle', in Day, I (ed) *Over a Red-Hot Stove: Essays in Early Cooking Technology.* Leeds Symposium on Food History 'Food and Society' Series. Totnes: Prospect Books, 83–98

Brown, S C 1979 *Benjamin Thompson, Count Rumford.* Cambridge (MA): MIT Press

Burgess-Wise, D 1977 *Fire Engines and Fire-Fighting.* London: Octopus Books

Campbell, S 1998 'Glasshouses and Frames, 1600–1900', *in* Wilson, C A (ed) *The Country House Kitchen Garden 1600–1950: How Produce was Grown and How it was Used.* Stroud: Sutton Publishing in association with the National Trust, 100–114

Carr, N and Gurney, I 1996 *Waddesdon's Golden Years, 1874–1935.* Stroud: Sutton Publishing

Carson, C 2009 *Technology and the Big House in Ireland c 1800–c 1930.* Amherst: Cambria Press

Cavendish, W G S (6th Duke of Devonshire) 1845 *Handbook to Chatsworth and Hardwick.* Privately printed

Cecil, Lady G 1931 *Life of Robert, Marquis of Salisbury, Vol III.* London: Hodder & Stoughton

Cecil, Lord D 1951 *Hatfield House: An Illustrated Survey of the Hertfordshire Home of the Cecil Family.* Derby: English Life Publications

Charlton, L E O (ed) 1949 *The Recollections of a Northumbrian Lady 1815–1866.* London: Jonathan Cape

Chartered Institution of Building Services Engineers 1997 *The Quest for Comfort.* London: Chartered Institution of Building Services Engineers

Christie, C 2000 *The British Country House in the 18th Century.* Manchester: Manchester University Press

Cole, G D H 2010 *The Life of William Cobbett.* London: Taylor & Francis

Colvin, H, Brown, R A and Taylor, A J P 1963 *History of the King's Works, Vol I: The Middle Ages.* London: HMSO

Combe, A 1842 *Principles of Physiology Applied to the Preservation of Health* (11th edition). Edinburgh: Maclachlan, Stewart & Co (first published 1834)

Cooper, Lady D 1958 *The Rainbow Comes and Goes.* London: Rupert Hart-Davis

Cox, R 1984 *Oh, Captain Shaw: The Life Story of the First and Most Famous Chief of the London Fire Brigade*. London: Victor Green Publications

Craven, M 1996 *John Whitehurst of Derby: Clockmaker and Scientist*. Ashbourne: Mayfield Books

Craven, M and Stanley, M 1991 *The Derbyshire Country House*. Derby: Breedon Books

Day, I 2009 'The Clockwork Cook', *in* Day, I (ed) *Over a Red-Hot Stove: Essays in Early Cooking Technology*. Leeds Symposium on Food History 'Food and Society' Series. Totnes: Prospect Books, 99–124

Dickson, P G M 1960 *The Sun Insurance Office 1710–1960*. London: Oxford University Press

Dillon, M 2002 *Artificial Sunshine: A Social History of Domestic Lighting*. London: The National Trust

Dillon, M 2012 'Advances in Lighting Technology and the Transformation of the Domestic Interior: A Case Study of Knole, Sevenoaks, Kent', *in* Barnwell, P S and Palmer, M (eds) *Country House Technology*. Donington (Lincolnshire): Shaun Tyas, 93–107

Drake and Gorham Ltd 1905 *Light and Power: A Treatise on the Application of Electric Current to Every Phase of Country Houses and Estate Requirements*. London: Drake and Gorham

Du Prey, P de la R 1987 'Eight Maids A-Milking'. *Country Life* 5 March 1987, 120–2

Eassie, W 1872 *Healthy Houses: A Handbook*. London: Simpkins, Marshall & Co

Eaton Estate 2008 *The Eaton Railway*. Chester: Grosvenor Estate

Ekirch, A R 2005 *At Day's Close: A History of Nighttime*. London: Weidenfeld and Nicolson

Evans, S 2011 *Life Below Stairs in the Victorian and Edwardian Country House*. London: National Trust

Eveleigh, D 1983 *Firegrates and Kitchen Ranges*. Princes Risborough: Shire Publications

Eveleigh, D J 2002 *Bogs, Baths and Basins: The Story of Domestic Sanitation*. Stroud: Sutton Publishing

Eveleigh, D J 2011 *Privies and Water Closets*. Princes Risborough: Shire Publications

Falkus, M E 1967 'The British Gas Industry before 1850'. *Economic History Review*, 2nd series, XX, 494–508

Fletcher, B F and Fletcher, H P 1910 *The English Home*. London: Methuen

Foulds, M 2011 *The Gorbals Brass and Bell Foundry – Bellfounding in Victorian and Edwardian Glasgow*. Dunblane: The Whiting Society of Ringers

Franklin, B 1806 *The Complete Works in Philosophy, Politics, and Morals, of the Late Dr Benjamin Franklin, Now First Collected and Arranged: With Memories of his Early Life* (Vol II). London: Longman, Hurst, Rees and Orme

Franklin, J 1975 '"Troops of Servants": Labour and Planning in the Country House 1849–1914'. *Victorian Studies*, 19/2, 211–239

Franklin, J 1981 *The Gentleman's Country House and its Plan 1835–1914*. London: Routledge & Kegan Paul

Fyfe, F 2012 'Harnessing Heat in the Kitchen Garden', in Barnwell, P S and Palmer, M (eds) *Country House Technology*. Donington (Lincolnshire): Shaun Tyas, 154–75

Garnett, O 2001 *Canons Ashby*. London: National Trust (Enterprises) Ltd

Gaskell, E 1855 *North and South*. (Reprinted by Penguin Books, London, 1994)

Gerard, J 1994 *Country House Life: Family and Servants 1815–1914*. Oxford: Blackwell

Gifford, A 2011 'George Sorocold – The Forgotten Water Engineer'. *Industrial Heritage*, 35(3), 2–17

Girouard, M 1978 *Life in the English Country House: A Social and Architectural History*. New Haven: Yale University Press

Girouard, M 1979 *The Victorian Country House* (2nd edn). New Haven and London: Yale University Press

Gladden, R 2011 *The Crewes of Crewe Hall*. Nantwich: Jerry Hall

Glover, S 1830 *The Peak Guide*. Derby: Henry Mozely and Son

Glynn, J 1853 *Rudimentary Treatise on the Power of Water*. London: John Weale

Gooday, G 2008 *Domesticating Electricity: Technology, Uncertainty and Gender, 1880–1914*. London: Pickering & Chatto

Gordon, D G 1911 'Refrigeration', *in* Weaver, L (ed) *The House and Its Equipment*. London: Country Life and George Newnes; New York: Charles Scribner's Sons, 93–96

Gordon, J E H 1891 *Decorative Electricity*. London: Sampson, Low, Marston, Searle & Rivington

Grant, F 2013 *Glasshouses*. Oxford: Shire Publications

Gray, L E 2002 *From Ascending Rooms to Express Elevators: A History of the Passenger Elevator in the 19th Century*. Mobile: Elevator World

Gregory, R and Miller, I 2013 *Uncovering the Estate: The Archaeology of Dunham Massey*. Lancaster: Oxford Archaeology Ltd

Hacker, C L 1960 'William Strutt of Derby (1756–1830)'. *Journal of the Derbyshire Archaeological and Natural History Society*, 80, 49–70

Hall, E 1989 'Hot Walls: An Investigation of their Construction in some Northern Kitchen Gardens'. *Garden History* 17.1, 95–107

Hann, A 2013 'Osborne Calling'. *Stories of England, English Heritage Membership Magazine*, July 2013, 31–2

Hardyment, C 1988 *From Mangle to Microwave: The Mechanisation of Household Work*. Cambridge: Polity Press

Hardyment, C 1992 *Home Comforts: A History of Domestic Arrangements*. London: Viking in association with the National Trust

Hare, A 1890 *The Story of My Life, Vol VI*. London: George Allen

Hartcup, A 1980 *Below Stairs in the Great Country Houses*. London: Sidgwick & Jackson

Hellyer, S Stevens 1891 *Principles and Practice of Plumbing*. London: George Bell & Sons

Hennessey, R A S 1972 *The Electric Revolution*. Newcastle upon Tyne: Oriel Press

Hewlings, R 2013 'The Dairy at Kenwood'. *English Heritage Historical Review*, 8, 36–81

Hird, M 1911 'Review of Lighting Systems', *in* Weaver, L (ed) *The House and Its Equipment*. London: Country Life and George Newnes; New York: Charles Scribner's Sons, 104–6

Hiskey, C 2007 'Palladian and Practical: Country House Technology at Holkham Hall'. *Construction History*, 22, 3–25

Hiskey, C 2011 'Holkham's Walled Gardens'. *Norfolk Gardens Trust Journal*, Spring 2011, 41–51

Hiskey, C 2012 'Palladian and Practical: Country House Technology at Holkham Hall', *in* Barnwell, P and Palmer, M (eds) *Country House Technology*. Donington (Lincolnshire): Shaun Tyas, 22–36

Holden, P 2007 *Lanhydrock*. Swindon: National Trust

Holden, P 2012 '"Is it Scientific and Safe?": Country House Technology at Lanhydrock in Cornwall', *in* Barnwell, P S and Palmer, M (eds) *Country House Technology*. Donington (Lincolnshire): Shaun Tyas, 58–74

Holt, R 1988 *The Mills of Medieval England*. Oxford and New York: Basil Blackwell

Hopwood, R 2004 *Fountains and Water Features*. Princes Risborough: Shire Publications

Hunwick, C 2012 'A Dynamo for a Duke: Hydroelectric Power at Alnwick Castle', *in* Barnwell, P S and Palmer, M (eds) *Country House Technology*. Donington (Lincolnshire): Shaun Tyas, 124–35

Irlam, G 2009 *The Hydraulic Lift at Dunham Massey*. Unpublished report for English Heritage

Jacks, L 1881 *The Great Houses of Nottinghamshire and the County Families*. Nottingham: W & A S Bradshaw

Keith, J 1989 *In Search of Old Gasworks*. Unpublished manuscript, Institution of Gas Engineers, Scottish Section

Kerr, R 1864 *The Gentleman's House, or How to Plan English Residences, from the Parsonage to the Palace, with Tables of Accommodation and Cost, and a Series of Selected Plans*. London: J Murray

Knight, J H 1906 *Electric Light for Country Houses*. London: Crosby, Lockwood & Co

Lawton, B 2012 'Tunnelling in the 1740s: The Water Mine at Coleshill House'. *International Journal for the History of Engineering and Technology*, 82(2), 187–209

Leatham, V 1992 *Burghley: The Life of a Great House*. London: The Herbert Press

Leeds City Art Galleries and Rutherford, J 1992 *Country House Lighting 1660–1890* (Temple Newsam Country House Studies No 4). Leeds: Leeds City Art Galleries

Leleux, S A 2014 'Railways at Belton House'. *The Narrow Gauge*, 224, 12–15

Lemmoin-Cannon, H 1912 *A Textbook on Sewage Disposal in the United Kingdom*. London: St Bride's Press

Lewis, M J T 2003 'Bar to Fish-Belly: The Evolution of the Cast-Iron Edge Rail', in Lewis, M J T (ed) *Early Railways 2: Papers from the Second Early Railways Conference*. London: The Newcomen Society

Lockhart, J G 1844 *Memoirs of the Life of Sir Walter Scott (new edition, complete in one volume)*. Edinburgh: Robert Cadell

Loudon, J C 1806 *A Treatise on Forming, Improving and Managing Country Residences*. London: Longman, Hurst, Rees and Orme

Loudon J C 1830 *The Gardener's Magazine and Register of Rural and Domestic Improvement, Volume VI*

Loudon, J C 1835 *An Encyclopaedia of Gardening* (first published 1822). London: Longman, Rees, Orme, Brown, Green and Longman. (Facsimile edition Garland Publishing, New York 1982)

Loudon, J C 1846 *Encyclopaedia of Cottage, Farm and Villa Architecture* (2 volumes) (first published 1833). (Facsimile edition Donhead Publishing, Shaftesbury 2010)

McGarry, D 1988 *Historic Communications Systems for Buildings, 1750–1910*. Unpublished Dip Arch thesis, The Architectural Association School of Architecture

McKenzie, P 1983 *W G Armstrong – A Biography*. Newcastle upon Tyne: self-published

MacNeill, I 1972 *Hydraulic Power*. London: Longman

Malcolmson, P E 1986 *English Laundresses: A Social History 1850–1930*. Urbana and Chicago: University of Illinois Press

Mander, P 1997 *The Fall and Rise of the Stately Home*. New Haven and London: Yale University Press

Margary, H Y 1905 Bells and Bell-Hanging, in Middleton, G A T (ed) *Modern Buildings, Their Planning, Construction and Equipment, Vol II*. London: Caxton Publishing Company, 191–5

Marshall, G, Palmer, M and Neaverson, P A 1992 'The History and Archaeology of the Calke Abbey Limeyards'. *Industrial Archaeology Review*, 14.2, 145–76

Marshall, P 1996 *Wollaton Hall: An Archaeological Survey*. Nottingham: Nottingham Civic Society

Martin, T and Seeley, N 2006 'Historic House Technology', in *The National Trust Manual of Housekeeping*. Oxford: Elsevier Butterworth-Heinemann, 661–9

Matthews, W 1835 *Hydraulia: An historical and descriptive account of the water works of London and the contrivances for supplying other great cities in different ages and countries*. London: Simkin, Marshall & Co

Mayhew, A and Mayhew, H 1847 *The Greatest Plague of Life: or The Adventures of a Lady in Search of a Good Servant*. London: David Bogue

Menuge, A 1993 'The Cotton Mills of the Derbyshire Derwent and its Tributaries'. *Industrial Archaeology Review*, XVI(1), 38–61

Merryweather, J C 1884 *Fire Protection in Mansions*. London: Merritt & Hatcher

Minnis, J 2009 *Sir David Salomons' Motor Stables, Broomhill, Southborough, Tunbridge Wells, Kent*. London: English Heritage Research Department Report Series, No 7-2009

Morris, C (ed) 1982 *The Illustrated Journeys of Celia Fiennes c 1682–c 1712*. London: MacDonald & Co

Morrison, K A and Minnis, J 2012 *Carscapes: The Motor Car, Architecture and Landscape in England*. New Haven and London: Yale University Press for the Paul Mellon Centre for Studies in British Art in association with English Heritage

Morritt, H E 1929 *Fishing Ways and Wiles*. Boston and New York: Houghton Mifflin

Moss, M 2002 *The Magnificent Castle of Culzean and the Kennedy Family*. Edinburgh: Edinburgh University Press

Munby, J 2008 'From Carriage to Coach: What Happened?', in Bork, R and Kaan, A (eds) *The Art and Science of Medieval Travel*. AVISTA Studies in History of Medieval Technology, Science and Art. Farnham: Ashgate Publishing, 41–54

Musson, J 2009 *Up and Down Stairs: The History of the Country House Servant*. London: John Murray

Muthesius, H 1904 *The English House*. Berlin: Wasmuth; first English edition 1979, London: Crosby Lockwood Staples

Newby, B 2000 'St Michael's Mount Tramway'. *Journal of the Trevithick Society*, 27, 18–24

Niblett, J T 1893 *A Popular Treatise on Portative Electricity*. London: Biggs & Co

Nicholson, C 2009 *The National Trust Carriage Museum at Arlington Court (revised 2011)*. National Trust: www.nationaltrust.org.uk/document-1355766877119/

O'Connor, H 1909 *Petrol Air-Gas*. London: Crosby, Lockwood & Son

O'Dea, W 1958 *The Social History of Lighting*. London: Routledge & Kegan Paul

Palmer, M 2012, 'Mills, Mines and Furnaces: Industrial Development and Landed Estates', in Barnwell, P S and Palmer, M (eds) *Country House Technology*. Donington (Lincolnshire): Shaun Tyas, 195–203

Palmer, M and West, I 2013 'Comfort and Convenience at Audley End House'. *English Heritage Historical Review*, 8, 82–103

Palmer, M and West, I 2013a 'Research into the Country House Technology at Audley End House, Essex.' Unpublished report for English Heritage

Palmer, M and West, I 2014 *Research into the Country House Technology at Eltham Palace, London SE9*. Unpublished report for English Heritage

Patrick, J 2012 *Walter Cave: Arts and Crafts to Edwardian Splendour*. Andover: Phillimore

Perkins, A M 1840 *Improved Patent Apparatus for Warming and Ventilating Buildings*. London: J B Nichols & Son

Phillips, R 1920 *The Servantless House*. London: Country Life

Pitt, W 1817 *A Topographical History of Staffordshire*. Newcastle-under-Lyme: J Smith

Price, A 1995 *The Building of Woodchester Mansion*. Stonehouse: Woodchester Mansion Trust

Raffald, E 1806 *The Experienced English Housekeeper* (13th edition). London: R Baldwin

Richardson, C J A 1837 *A Popular Treatise on the Warming and Ventilation of Buildings*. London: John Weale

Roberts, B M 2008 *Historic Building Systems and Equipment: Heating and Ventilation*. London: English Heritage

Roberts, J and Hargreaves, M 2002 'Stephen Switzer, Hydrostaticks and technology in the Country House Landscape'. *Transactions of the Newcomen Society*, 73, 163–78

Roberts, R 1827 *The House Servants' Directory, or A Monitor for Private Families Comprising Hints on the Arrangement and Performance of Servants' Work*. Boston: Munroe and Francis

Robinson, J M 1983 *Georgian Model Farms*. Oxford: Clarendon Press

Robinson, J M 2011 *Felling the Ancient Oaks: How England Lost its Great Country Estates*. London: Aurum Press

Robinson, J M 2012 *James Wyatt (1746–1813): Architect to George III*. New Haven: Yale University Press

Rowell, C 2006 'The Historic House Context – the National Trust experience', in *The National Trust Manual of Housekeeping*. Oxford: Elsevier Butterworth-Heinemann, 9–19

Sage, G 2014 'Carstairs House Tramway'. *The Narrow Gauge*, 229, 8–11

Salomons, D 1890 *Electric Light Installations and the Management of Accumulators*. London: Whittaker & Co

Sambrook, P A 1989 *A Servant's Place*. Stafford: Shugborough Estate

Sambrook, P A 1996 *Country House Brewing in England, 1500–1900*. London: Hambledon Press

Sambrook, P A 1999 *The Country House Servant*. Stroud: Sutton Publishing

Sambrook, P A 2003 *A Country House at Work: Three Centuries of Dunham Massey*. London: National Trust Enterprises

Sambrook, P A 2012 '"The Servants' Friend?" Country House Servants' Engagement with New Technology', *in* Barnwell, P S and Palmer, M (eds) *Country House Technology*. Donington (Lincolnshire): Shaun Tyas, 75–92

Sambrook, P A and Brears, P 1997 *The Country House Kitchen 1650–1900*. London: National Trust Enterprises in association with Sutton Publishing Ltd (paperback edn)

Scott, A A H 1911 'Sewage Disposal', *in* Weaver, L (ed) *The House and Its Equipment*. London: Country Life and George Newnes; New York: Charles Scribner's Sons, 93–6

Sekers, S 1998 'The Walled Gardens at Shugborough in Staffordshire', in *The Country House Kitchen Garden 1600–1950: How Produce was Grown and How it was Used*. Stroud: Sutton Publishing in association with the National Trust, 62–99

Seligman, S (ed) 2008 *Explore the Gardens at Chatsworth*. Chatsworth: Chatsworth House Trust

Semmens, P W B 1964 'St Michael's Mount Tramway'. *The Railway Magazine*, 101 (July 1964), 585–8

Shifrin, M 2015 *Victorian Turkish Baths*. Swindon: Historic England

Shröder, M 1969 *The Argand Burner: Its Origin and Development in France and England 1780–1800*. Odense: Odense University Press

Shuffrey, L A 1912 *The English Fireplace*. London: Batsford

Singleton, A 2009 'Canwick Hall Sewage Treatment Plant'. *Lincolnshire History and Archaeology*, 44, 42–5

Smith, P 2000 'Welbeck Abbey and the 5th Duke of Portland: Eccentricity or Philanthropy', *in* Airs, M (ed) *The Victorian Great House*. Oxford: Oxford University Department of Continuing Education

Smith, P 2010 *The Motor Car and the Country House*. English Heritage Research Department Report Series, No 94-2010, http://services.english-heritage.org.uk/ResearchReportsPdfs/094_2010WEB.pdf

Smith, P 2012 'Wollaton Hall: Technology and the Regency Country House', *in* Barnwell, P S and Palmer, M (eds) *Country House Technology*. Donington (Lincolnshire): Shaun Tyas, 37–57

Smith, T 2003 *Hydraulic Lifts at Patshull Hall, version 4*. Unpublished report produced for the owner

Squires, S 2012 'Country House Tramways: Belton House, Harlaxton Manor and Stoke Rochford Hall'. *Lincolnshire History and Archaeology*, 47, 29–34

Staniforth, S 2006 'Conservation: Principles, Practice and Ethics', in *The National Trust Manual of Housekeeping*. Oxford: Elsevier Butterworth-Heinemann, 34–43

Stevenson, J J 1880 *House Architecture*, Vol II. London: Macmillan & Co

Sutcliffe, G L 1907 *The Modern Plumber and Sanitary Engineer*, Vol IV. London: Gresham Publishing Company

Swan, M E and Swan, K R 1968 *Sir Joseph Swan FRS: Inventor and Scientist*. Newcastle upon Tyne: Oriel Press (first published 1929 by Ernest Benn Ltd)

Switzer, S 1729 An *Introduction to a General System of Hydrostaticks and Hydraulicks, Philosophical and Practical*. London: T Astley

Sylvester, C 1819 *The Philosophy of Domestic Economy*. London: Longman, Hurst, Rees, Orme and Brown

Tann, J 1970 *The Development of the Factory*. London: Cornmarket Press

Thompson, F M L 1963 *English Landed Society in the 19th Century*. London: Routledge & Kegan Paul

Thynne, D W L (Marchioness of Bath) 1951 *Before the Sunset Fades*. Longleat: Longleat Estate Co

Tomory, L 2012 *Progressive Enlightenment: The Origins of the Gaslight Industry, 1780–1820*. Cambridge (MA): MIT Press

Turberville, A S 1939 *A History of Welbeck Abbey and its Owners*. Vol II, 1785–1879. London: Faber and Faber

Turvey, R 1993–4 'London Lifts and Hydraulic Power'. *Transactions of the Newcomen Society*, 65, 147–64

Ure, A 1835 *The Philosophy of Manufactures*. (Facsimile edition Frank Cass & Co, London, 1967)

Vries, L de 1991 *Victorian Inventions*. London: John Murray

Wade-Martins, S 1991 *Historic Farm Buildings*. London: Batsford

Wade-Martins, S 2002 *The English Model Farm: Building the Agricultural Ideal, 1700–1914*. Macclesfield: Windgather Press in association with English Heritage

Walford, E 1875 *Walford's County Families of the United Kingdom*. London: Robert Hardwicke

Walker, H 1934 *Reminiscences of 70 Years in the Lift Industry*. Self-published

Walker, W 1850 *Useful Hints on Ventilation*. Manchester: J T Parkes

Ward, J T and Wilson, R G 1971 *Land and Industry: The Landed Estate and the Industrial Revolution; A Symposium*. Newtown Abbot: David and Charles

Waterson, M 1980 *The Servants' Hall: A Domestic History of Erddig*. London: Routledge & Kegan Paul

Weaver, L (ed) 1911 *The House and Its Equipment*. London: Country Life and George Newnes; New York: Charles Scribner's Sons

Weightman, G 2002 *The Frozen Water Trade*. London: Harper Collins

Weightman, G 2011 *Children of Light: How Electricity Changed Britain Forever*. London: Atlantic Books

Weskill, O 1997 *Provincial Insurance Company 1903–38*. Manchester: Manchester University Press

West, I 2009 *Light Satanic Mills – The Impact of Artificial Lighting in Early Factories*. Unpublished PhD thesis. School of Archaeology and Ancient History, University of Leicester

West, I and Palmer, M 2013 *Technology at Tatton Park*. Unpublished report for Cheshire East Council

Whitbread, H (ed) 1992 *No Priest but Love: Excerpts from the Diaries of Anne Lister 1824–1826*. Otley: Smith Settle

White, R 1875 *Worksop, 'The Dukery' and Sherwood Forest*. London: Simpkins, Marshall & Co

Wilcock, J 1997 'On the Water Courses of Shugborough'. *Journal of the Staffordshire Industrial Archaeology Society*, 16, 20–5

Wilkinson, L P 1969 *The Georgics of Virgil: A Critical Survey*. Cambridge: Cambridge University Press

Wilson, P N 1957 'Early Water Turbines in the United Kingdom'. *Transactions of the Newcomen Society*, 31, 219–41

Wilson, R and Mackley, A 2000 *Creating Paradise: The Building of the English Country House 1660–1880*. London and New York: Hambleton and London

Wolfe, J J 1999 B*randy, Balloons and Lamps: Ami Argand, 1750–1803*. Carbondale and Edwardsville: Southern Illinois University Press

Worsley, G 2004 *The British Stable*. New Haven and London: Yale University Press for the Paul Mellon Centre for Studies in British Art

Wright, L 1960 *Clean and Decent: The Fascinating History of the Bathroom and Water Closet*. London: Routledge & Kegan Paul